# INTERTWINED

# INTERTWINED

## WOMEN, NATURE, AND CLIMATE JUSTICE

REBECCA KORMOS

NEW YORK
LONDON

A portion of the proceeds generated by this book will be donated to Planet Women, an organization that partners with women to create systemic change and invest in environmental solutions across the globe.

Published in the United States by The New Press, New York, 2024
Distributed by Two Rivers Distribution

ISBN 978-1-62097-749-1 (hc)
ISBN 978-1-62097-863-4 (ebook)
CIP data is available

The New Press publishes books that promote and enrich public discussion and understanding of the issues vital to our democracy and to a more equitable world. These books are made possible by the enthusiasm of our readers; the support of a committed group of donors, large and small; the collaboration of our many partners in the independent media and the not-for-profit sector; booksellers, who often hand-sell New Press books; librarians; and above all by our authors.

www.thenewpress.com

*Book design and composition by Bookbright Media*
*This book was set in Bembo and Kievit*

Printed in the United States of America

10 9 8 7 6 5 4 3 2 1

*For Tess, Chloe & Cyril,*

*all Earth's Children,*

*and the Earth Herself*

*You are welcome here.*

*You—as you—*

*Looking like you,*

*talking like you,*

*laughing like you,*

*moving the way that you move,*

*are welcome here.*

*This place is for you.*

*That's the hope.*

—Faith Briggs, *This Land*[1]

# CONTENTS

# LAND ACKNOWLEDGMENT

I RESEARCHED AND WROTE THIS BOOK WHILE LIVING ON THE TERRITORY of Xučyun (pronounced *Hoo-chi-oon*), the ancestral and unceded land of the Chochenyo Ohlone, the successors of the sovereign Verona Band of Alameda County. The land where I live was and continues to be of great importance to many tribal Ohlone Nations, who are alive and flourishing members of what is now colonially known as Berkeley. I acknowledge that I have benefited from living on this land and commit to actively pursuing values of anti-racism and anti-colonialism, as well as supporting and amplifying the voices of Indigenous peoples and tribes, in this book and in my life and work.

# PREFACE

THIS BOOK BEGAN AS A PERSONAL JOURNEY TO UNDERSTAND THE intertwined relationship between women, nature, and climate justice. Women across the planet are working fervently and passionately to protect animals, plants, oceans, rainforests, insects, and fungi, to restore nature, and to draw down greenhouse gasses from the earth's atmosphere. At the same time, women are unequally impacted by the climate crisis—floods, droughts, and extreme temperatures overwhelmingly and disproportionately affect women in the short and long term. Despite this disparity women do not have an equal voice in decision-making about climate change or nature's protection. Yet when women are included equally, the results for women, their communities, and the environment are tremendous. This entanglement between women and nature—whether borne out of passion, necessity, or even violence, lies at the heart of this book.

I should have been born into the body of a wild animal. I was a wild child—or perhaps I just did not control the wild that exists within us all. My dreams were of the wilderness, tropical rainforests, and living among communities of wild animals. I did not fit well into the confines of city life. Early on, almost from birth, chimpanzees held a particular fascination for me. They were so like humans, and yet not like us as well. I wanted to understand where those similarities and differences began and ended.

When I was twenty-four, I traveled to live in the rainforests of Gabon—to the Lopé National Park—to study a species of monkey called the gray-cheeked mangabey. It was a time in my life when everything seemed abundant. Every day, I walked under enormous trees with giant buttresses and extensive canopies whose beauty was expressed in every shade of green possible. The forest was a place where gorillas beat their chests and hummed as they fed on fruit high in a tree, where the blood-curdling screams of mandrill females harmonized with the rhythmic, raspy grunts of the males with purple and red bottoms as they moved in groups of over eight hundred monkeys, advancing like a massive, noisy demonstration. Elephants, with their slate-gray, thick-skinned masses, glided silently and gracefully along their centuries-old paths through dense vegetation between fruiting trees. Pythons, vipers, and black cobras slithered through the leaf litter on the forest floor. Giant fruit bats spread large, light-brown translucent wings, while leopards silently hunted the sleeping monkeys draped over branches in the canopy above. There were razor grasses that sliced your skin, rare soft orchids growing on moist chocolate-brown branches, tumbling rivers, and torrential rains. It was a world of wondrous diversity, fertility, richness, and plenty.

My elation of living in a rainforest was cut short, however, when after only six months in Gabon, I was attacked by a wounded buffalo dying in a streambed. His horn penetrated my breast, ankle, and hand before he left me bleeding in the river. I had joined in an intimate entanglement with a wild animal that left a part of him within me forever, yet in that moment a part of me was lost too. Feelings of invincibility that I carried with me in my youth floated away downstream with the intermingling of our cranberry-red blood. In that moment, the extinction of life became something incredibly real, personal, and possible.

After months of healing in Canada, then Scotland, I returned to Gabon. Although my physical wounds had healed, the emotional scars went much deeper. Fear infected me such that every rustling leaf or raindrop was a buffalo coming to spear me again. On the verge of quitting, I hired an old hunter, Francis Nzinga, from the Lopé village to accompany me in the forest. Over the next few years, we walked together almost every day. Through his gentle teachings, I learned about the medicinal properties of plants, how to read the forest's record of which animal had passed by and when. I learned to call a forest antelope by whistling through a leaf, to make a cup for drinking river water by folding a large flat *Marantaceae* leaf, to smell a gorilla before I saw one, and to feel an elephant rumble before I heard it. I learned how Francis's life, and that of his community, was integrally connected to the forests, wildlife, and the river.

I lived in a research station of foreign researchers. Our methods were scientific. We used compasses, we took notes, and we used statistics to demonstrate relationships among the animals. Francis didn't need a compass, and he understood that relationships were lived.

After three years, I left Gabon for Guinea, West Africa. In Guinea, it was as if I were walking into the future. More than 85 percent of Guinea's native vegetation cover was already gone. The abundance of life I found in Gabon was no longer something to be taken for granted. Extinction expanded to become something that I not only feared but could see, hear, and smell. Extinction was an empty forest eerily quiet without monkey or bird song, a path unbroken by hoofprints, and air without the smell of musky wet fur or fresh dung. Desertification was the soot in my nose, the silt in the streams, and the heat that rose from ground unshaded as I walked across it.

I was in Guinea to survey chimpanzees because, even by the mid-1990s, they had disappeared from many parts of the country. In a mountainous area of Guinea called the Fouta Djallon, however, there were more chimpanzees than anywhere in West Africa. The reason for their continued survival was explained through a legend recounting how chimpanzees used to be humans. One day, the humans went against the wishes of God, and they were banished to the forest, where they became wild. Because they are our ancestors, people did not hunt them. The chimpanzees and the communities' lives were closely intertwined. The continued existence of the chimpanzees was not in spite of people—but because of them.

Almost twenty years later, I returned to Guinea with a friend and filmmaker, Kalyanee Mam, to work with the nongovernmental organization Guinée Écologie to make a documentary film about the relationship between people and chimpanzees. As I had done throughout my years of research, I prepared to interview hunters and heads of villages—roles typically reserved for men. Instead, Kalyanee suggested that we expand to include and highlight women's stories as well. This is how I met N'yamnama Traoré.

When we asked N'yamnama if we could talk with her about the chimpanzees, I was surprised to hear her say, "Nobody ever asked me what I think before." As N'yamnama spoke about growing up with the chimpanzees in the mountains surrounding Bossou, she told us a story about when she was a girl. Each year, just before harvest, her family would make an offering to the chimpanzees. "Here is your part. Eat. And what remains in the field—that is for my children who tend the crops," her uncle used to say, as he placed bananas at the edge of the field.[1] But when the sacred mountain fell under the management of the state, N'yamnama and her family were displaced, and the people's and chimpanzees' lives became separated. Over time, the forests of Bossou became degraded, and the corridor of trees between Bossou and nearby forests was cut down. In their isolation, the chimpanzees have become more vulnerable, especially to disease. When I visited this same mountain in the mid-1990s, there were about twenty-four chimpanzees. By the end of 2023, there were only four left.

In another village far to the north in Guinea, another woman spoke of conflicts with chimpanzees over water and how these were more frequent when there was less rain. The women didn't go alone to gather water but traveled in groups. If there were chimpanzees there, they waited—each taking their turns at the source. These were things I did not hear two decades earlier: concerns of water scarcity and human-wildlife conflict, but also stories of reciprocity and sharing. I did not hear the full story, because I had not been listening to half the population.

For most of my life as an environmentalist, I have worked principally with men. I have conducted my projects in what is called a "gender-blind" manner—that is, not considering the gendered roles and responsibilities that society ascribes to or imposes upon a person. When I worked with hunters and fishermen like Francis, or interviewed heads of villages as they talked about their land and their home, I understood a piece of the picture but not the whole thing. I thought that ignoring gender put everyone on an equal footing. It does not. By not considering gender, I was reinforcing those very inequalities I wanted to eliminate.

What I didn't understand is that gender equality cannot be addressed independently of protecting nature. Nor can protecting nature ever be truly sustainable without addressing gender inequality. Gender equality is not a *consequence* of a restored planet. It is not something to be addressed *after* we manage to slow greenhouse gas emissions. The rights of women and the protection of nature go hand in hand with a sustainable future—they are integrally connected.

When N'yamnama said how happy she was to be interviewed and that no one had asked for her opinion before, I felt ashamed. Why had I not made the effort to speak with women? Why had I almost always chosen to hire and collaborate with men?

Driven to understand the reasons for my biases, I started listening. I began participating in initiatives and forums organized by women for women. I was shocked to learn about the disproportionate impacts of climate change on women. I also learned how women were not being equally included in the management of natural resources or in decisions about nature—from grassroots conservation projects to regional initiatives to policy-making forums about climate change.

When I investigated what was being done to address these social inequities, I learned that women across the planet have been joining hands and raising their voices, forming coalitions, building networks, and creating an emboldened women's environmental movement—and that this movement was brave, fierce, and gathering momentum. At its core, it has two crucial tenets. The first is the understanding that protecting the environment requires advancing women's rights, because degradation of nature and social injustices toward women are inextricably linked. The second is that we must empower women because their leadership capacities and knowledge can steer us away from our current disastrous trajectory toward a dying planet and the sixth extinction crisis.

I had so many questions. What is this special knowledge that women hold about biodiversity and nature? Do women lead differently than men? Are there different results when women lead? I started searching for a book that might answer some of these questions. As a scientist, I was keen to look at the evidence.

But it was not just empirical evidence I was searching for. I sought something incredibly personal as well. I have always drawn my energy, happiness, and grounding from nature. In my younger years, my cycles were synchronized with the moon. Giving birth to my daughters was deeply beautiful and primal. I was swept up in the overwhelming passions of motherhood. Did being female, therefore, somehow make me closer to nature? Why do we call our planet "Mother Earth," and the wild "Mother Nature"? Was this gendering of the earth and nature appropriate, or could it be somehow damaging to women because it essentialized us? Does a feminized earth ignore Indigenous two-spirit people and create binaries where there are none? Does the personification of the earth as a woman ignore men, distance men from the earth, or even justify male domination of nature and the earth? I wanted a book that could both encourage and inspire me by exploring the historical, mythological, cultural, poetic, and artistic—as well as practical and ecological—relationships women have with nature. In summary, I wanted to understand better this powerfully entwined relationship between the two.

Although I did find information about women and climate change, I did not find a book that revealed the linkages between women and biodiversity loss as well, nor about how all these crises are intertwined. What information did exist was in scientific publications, white papers, or webinars. Some answers were available, but they were scattered across the web and difficult to find. It was when I started talking with women leaders of grassroots projects to protect and restore nature that I realized so much of the knowledge of women lives in our experience but has not yet been documented. Advice is often passed on verbally and only between each other.

This is how this book came to be. I decided that I would research and write the book that I so badly wanted to read. My aim was to bring together dispersed information, but I also sought to interview women to document an oral history. Over the past year, I have interviewed over fifty women to learn from the infinite wisdom that lives in our experience. I listened to women leaders in conservation nonprofits, women in governments, women philanthropists, and women activists, but most of all, I wanted to reach those women from whom we often do not hear enough—women on the front lines of climate change, women leading grassroots movements to protect biodiversity, and women working at the community level to empower other women as well as protect our planet.

Climate change is affecting the air around us, the water we drink, the wind, the rain, our food, and ultimately our security and safety as conflicts over scarce resources erupt. Those who realize how catastrophic our future could be have the same question on their lips: "What is it that we need to do?" Scientists have been very clear in their answer. We need to cut current global emissions in half by 2030, and then in half again by 2040, and then we need to cut emissions to net zero by 2050—"at the very latest."[2]

Luckily, many of our most brilliant minds have been offering solutions that make net zero by 2050 possible. But as I scour the literature, websites, and media articles, there is one crucial piece of the plan that has, in my view, not received ample attention. It is not a technological innovation, such as one that sucks $CO_2$ out of the atmosphere, nor is it a trade-off, such as an offset that permits harmful behaviors in one place with the promise of protection in another. In front of us all along has been the most simple, beautiful, and powerful way to bend our trajectory regarding current climate change and species loss, but like so many other things related to women, it has been overlooked, erased, and excluded. Not only does it redirect us to a more sustainable and healthy future, it also addresses issues of social injustice—what some might call a win-win solution.

The idea is this: empower women and the world will change. This is not about excluding anyone; it is about valuing everyone equally. It is about rising as one to the challenges of climate change and biodiversity loss. I will delve into what I mean by "empowerment" later, but in a nutshell, it refers to all the efforts by which we can achieve gender equality at a global scale. Critically, this involves equal access to education, equal rights to own land and earn money, equal decision-making power, and—crucially—reproductive choice. Specific to the environment, empowerment involves equal representation of women in decision-making bodies that govern and manage natural resources, and the structuring of social and economic systems so that women can participate equally in decision-making, management, and protection of nature.

A plethora of studies now show that when women are empowered in these ways, remarkable and rippling results ensue. Gender equality is linked to better health, happiness, and nutrition for not only women themselves but their families and communities too. In fact, women's leadership and/or equal representation is linked to lower $CO_2$ emissions, better forest management, better land protection, less land-grabbing, fewer conflicts over resources, and better compliance with rules governing those resources. I hope that after reading this book, you will be convinced as well that women's empowerment and gender equality sits at the nexus of nearly everything. It is the solution that can provide the fertile ground for humanity and all life on earth to have a healthy, sustainable, and resilient future.

When N'yamnama speaks about the chimpanzees from her childhood, she expresses worry that there are so few remaining. "For us, they are our soul," she says. Through our desire for "development," humans have been separated from nature, and from each other. We have lost our understanding of the importance of community and continuums. Thus, this book is ultimately about reconnecting. It is about recognizing the value of each individual and acknowledging our connection with the earth and one another.

# INTRODUCTION

It is February, 2023, and I receive a WhatsApp message from Madina Amin Hussein, Indigenous conservationist and passionate advocate for nature and women in northern Kenya, where droughts have plagued the region for five consecutive seasons: "I wanted to share with you the situation in our region. Drought has hit people hard. We need to raise funds so that we can buy foodstuffs. It's a matter of life and death." She continues, "The worst affected are the women. The drought has further heightened the risk of gender-based violence, sexual exploitation, and abuse. It is hampering girls' access to education."

Madina is the founder and managing director of Global Nature Conservation, a nonprofit based in Kenya's Garissa County—one of the regions most impacted by drought. I don't think I have ever seen such fiery determination in a conservationist. Madina is not only helping with relief efforts for people displaced by the drought, but is also working to reforest Garissa County, planting *Acacia tortilis*, a species of tree with an umbrella-shaped canopy, for people, reticulated giraffe, and other wildlife endemic to the region.

When she next travels to Nairobi, we speak on Zoom as her two-year-old nephew climbs in and out of her lap, peering into the computer. We laugh and talk about how women so often must work this way, especially during the pandemic, looking after young ones while trying to focus. But the conversation becomes more serious when Madina talks about her childhood. "I grew up in Bodhei forest," she says. "It is one of the biggest natural forests in the country. As I grew, I saw nature as my immediate family. I saw trees and wildlife as my relatives. But I came to learn that some of the trees that used to exist are no longer there. They have been destroyed by people for charcoal." In telling me this, she stops to wipe tears away from behind her glasses. "Sorry," she says, but I tell her that it is okay. One of the things I am learning is that we need to bring our feelings back into this work. We have shut them out for far too long in the name of professionalism. But this is emotional work.

Madina embodies so much with respect to the challenges and solutions that lie at the intersection of climate change, biodiversity loss, and gender. Her relationship with nature is driven by love, connection, and apprehension about its disappearance. She is bearing witness to the impact of climate change on forests and wildlife as well as humans, especially women. Her work is cross-sectoral, involving advocacy, policy, conservation, and humanitarian work all at once. The women who Madina supports are those who are most impacted by climate change yet contribute to it the least.

Women are disproportionately affected when nature is degraded, species are lost, and the climate changes. Women suffer greater mortality from climate-related fires, floods, and hurricanes, in some cases making up 90 percent of those killed by extreme weather events. In the chaos following these events, women continue to suffer greater mortality.[1] Because women are most impacted, it stands to reason that we should center and amplify women's wisdom, knowledge, stories, and solutions. Yet, women are underrepresented or excluded from almost every aspect of decision-making about nature and the future of our planet.

There is pervasive gender inequality within the leadership of the nonprofit conservation sector, where only about a quarter of CEOs were women in 2018.[2] The average level of representation of women in national and global climate negotiating bodies has hovered around one-third. We are not directing resources adequately to women to fix the problem either. Despite women's increased susceptibility to the climate and biodiversity crises, only about 0.2 percent of all foundation funding focuses intentionally on women and the environment.[3]

What would happen if these imbalances were redressed? As I dug into the research for this book and learned about the ripple effects of women leading and having equal rights, equal pay, equal education, and equal say, it became apparent that empowering women is singularly one of the most important solutions to climate change and biodiversity loss—and that fact is not being talked about enough. Empowering women benefits children, communities, countries, nature and climate at local, national, and global levels.

We are living in a time of great urgency to slow the emissions of greenhouse gases into the atmosphere. We do not have time on our side. My hope is that by bringing together currently scattered evidence from across disciplines to make a powerful case for the necessity of women's empowerment for addressing climate change and biodiversity loss, that we, as a global community, will shift the distribution of attention and funds to women and girls, and work to dismantle the barriers and oppressions to finally unleash women's full potential.

Humans frequently divide the world into binaries: right and wrong, positive and negative, north and south, winner and loser, civilized and wild, them and us, men and women. Once binaries are named, however clumsily, we breathe life into them, reinforcing their existence and ultimately using them to polarize. We begin to see the binaries as competing, conflicting, and opposing each other. We search for and repeat narratives to justify categories and reinforce the stereotypes that define them. When something or someone does not fit under a neat heading, we force them to fit anyway. In turn, the binaries we have created often force us to make false choices. Such binaries are most common in cis-heteronormative patriarchal cultures. Yet in some cultures, particularly Indigenous ones, there is greater fluidity between nature and humans, and between male and female.

I would like to highlight that when I use the word "women" here, it refers to cisgender women, transgender women, and AFAB (assigned female at birth), femme/feminine-identifying, two-spirit, genderqueer, nonbinary, and non-gender-conforming people. My focus on women is not to reinforce binaries or to exclude men. I focus on women because they are among the groups most affected by climate change and because they have historically and globally been excluded from much of the decision-making about the environment.

Even though it is becoming more broadly understood and accepted that gender is a continuum, the reality is that in many countries those deemed "female" are subject to controls, confines, constraints, exclusions, cultural expectations, pressures, and violence that those in the category of "male" never experience. Restrictions on what clothes women can wear, where and when they can travel, what jobs they can do, and what sort of resources, information, education, privileges, and freedoms they can have access to have been imposed on them, both historically and currently. Being a woman increases one's chances of experiencing violence from men. Being a woman means that you often do not have full control of what happens to your body. Central to this book is that being put in the category of "female" also results in being disproportionately impacted by the climate and biodiversity crises.

Because women are unequally affected, it is imperative that we focus on and amplify women's stories and solutions, and it is critical that this is done with an intersectional lens. Women are not a homogeneous group. We each have many intersecting identities that make up our whole, determining privilege or vulnerability, whether it be our gender, race, economic status, where we live in the world, or who our ancestors may have been.

Another false binary covered in this book is that between nature and culture. The Western worldview commonly separates humans from nature, but the world has not always been like this, and viewing humans as part of nature, instead of separated from it, is a common thread found in most Indigenous worldviews. The Western worldview, on the other hand, sees nature as falling into different tiers on a hierarchical pyramid, with humans at the top. The myth of humans as the pinnacle of creation is told in the illustration of human evolution that follows an ape/human ancestor through its progression into a man (almost never a woman). It positions man as the goal: the "natural" end point of successful evolution. But meeting a wild chimpanzee within their forest home feels nothing like meeting a less advanced creature.

The first time I saw a chimpanzee in the wild was in Gabon in 1990. The sound of fruit bits falling to the ground led me to a tree where a chimpanzee sat eating. The moment was brief. Upon seeing me, he dropped to the ground with a thud and ran off. It was not an encounter with a being less evolved than me, but rather a moment when I became fully aware of my shortcomings and weaknesses as a human. With thumbs on their feet as well as hands, chimpanzees had the ability to climb that I did not. They knew what forest foods they could eat, and how to fold branches to make a springing nest of soft vegetation high in a tree for sleeping. They had no need for shoes, clothing, a compass, or binoculars. In my boots and field pants, with my lunch in the pack on my back, I didn't feel superior. I felt clunky.

Connections go well beyond what we call the "living" or "animate," as well. We are also connected to the land, air, and oceans. As Robin Wall Kimmerer writes in *Braiding Sweetgrass*, "English doesn't give us many tools for incorporating respect for animacy. In English, you are either a human or a thing. Our grammar boxes us in by the choice of reducing a nonhuman being to an *it*, or *it* must be gendered, inappropriately, as a *he* or a *she*. Where are our words for the simple existence of another living being?"[4] As Midori Nicolson, a Dzawada̱'enux̱w First Nation fisheries and wildlife biologist, tells me, "Rocks have a life. All of these beings give and take."

I first became aware of Midori's work when I attended a Women in Nature Network webinar she presented titled "The Realm of the Sky: An Indigenous Perspective on Birds as Valuable Coastal Indicators." She spoke about the importance of birds in Kwakwaka'wakw culture and how they are often used as indicators of ecosystem health and species abundance. Midori somehow managed to seamlessly weave together two very different perspectives on nature—those of Western science and an Indigenous worldview. In a conversation we had several years after this webinar, Midori explains to me that it had not always been easy for her to merge the two ways of seeing:

> I come from a traditional teaching about who we are within our environment, totally different from Western science. We are already conservation biologists as Indigenous people. This is part of our training. This is part of our knowledge. But I struggled with the perspectives that I had been raised in understanding, and the more rigid, confined, analytical space of Western science. I can't quite unsee the way I was raised to look at the river or at the world and the creatures that surround us as equal or greater beings than ourselves. I call it the "quaintification" of our knowledge, but I don't see it as whimsical or quaint. People will only more recently see value and say, "Let's look at traditional knowledge. Let's find a way to put this Indigenous framework or

> this Indigenous knowledge into this other structure," but for those of us that have been staggering along with both those visions, it's hard to articulate where those boundaries are.

Now, more than ever, we need to be listening to the voices of those who have not lost a connection with nature. Today, more than one-quarter of land on our planet is owned, managed, used, or occupied by Indigenous peoples and local communities, and science is confirming what many Indigenous people knew all along—that these lands are among the best protected. A recent study presents new evidence that the level of protected nature in places managed by Indigenous people is equal to, or more than, that of places categorized as "strict protected areas" by the International Union for Conservation of Nature (IUCN).[5] In Bolivia, Brazil, and Colombia, rainforests managed by Indigenous communities have reduced deforestation and carbon emissions more than many areas that have a legal conservation status.[6] In addition, vertebrate species diversity is richer on many Indigenous-managed lands than in areas with legal protection. There is tragic irony in the fact that we are finally realizing that those people who historically have been excluded and marginalized hold the very wisdom the world needs to get out of the mess we are in.

There is one more binary in this book that I would like to address. Many studies distinguish wealthier countries from less wealthy ones, and many of the terms they use to describe that distinction have roots in racists and colonial pasts.[7] Historically, the terms "developing" and "developed" countries have been commonly used. This terminology, however, begs the question of whether unsustainable rampant destruction of our environment should be called "developed." Other terms, such as "first, second, and third world countries," have also been used as a way of dividing up the world. More recent terms have been "Global North" and "Global South," which designate where a country or region falls in relation to the "Brandt Line." This line places "richer" countries in the Northern Hemisphere (and, though not geographically accurate, the group includes Australia and New Zealand) and "poorer" countries in the Southern Hemisphere. These labels are all false dichotomies. Critically, they are problematic in that they ignore the differing poverty levels that people experience within countries. Wherever possible, therefore, in this book I commit to highlighting the diverse experiences of people within as well as between countries and across these global divisions.

I hope this book stretches across male-female, nature-culture, animate-inanimate, and north-south divides, and starts a journey toward bridging them. We need to relearn a way of speaking holistically about our commonalities and connections.

# PART I

# ROOTS

A tree has roots in the soil yet reaches to the sky. It tells us that in order to aspire we need to be grounded and that no matter how high we go it is from our roots that we draw sustenance.

—*Wangari Maathai*[1]

# 1

# THE PLANETARY POLYCRISIS

> Do some leaders in this world believe that they can survive and thrive on their own? Have they not learned from the pandemic? Can there be peace and prosperity if one-third of the world, literally, prospers and the other two-thirds of the world live under siege and face calamitous threats to our well-being?
>
> —*Mia Amor Mottley*[1]

THE TROPICAL FOREST OF LOPÉ, GABON, HAS ONE OF THE HIGHEST densities of elephants in the world. Back in the 1990s, as I walked each day searching for and following monkeys, this meant that I was running into elephants often. It was both wonderful and terrifying. Running into a group of elephants while you are on foot is a very different experience than seeing one from a car. Encounters with elephants could result in us quietly avoiding each other and continuing on our separate ways, but it could also mean a sudden trumpeting and crashing through the vegetation. After that, one of us would leave, usually at a sprint. Most often it was me.

In one swampy area that was particularly dense with vegetation, there was a giant treefall—a kapok tree, which is a relative of the massive baobab tree endemic to drier parts of Africa. The kapok's smooth, straight trunk must have once projected at least sixty meters toward the sky. Just as spectacular as its three-meter-thick trunk were the roots that had been ripped out of the earth when it fell. They were buttressed roots that extended about twenty meters into the air. The roots were twisted, entangled, woody, and easy to climb. When I heard elephants nearby, I would climb up to the top of this root system and have a rare opportunity to watch them safely from above.

The majority of a tree's roots are below the surface, and out of sight, but they are what provides water and nutrients to the whole system, what stabilizes it, and what grounds it. For me, it has always been those small moments of awe at the beauty of nature that have rooted and nourished me in a passion to protect it. Beauty lies in tiny seedlings growing side by side, sprouting from interlaced, threadlike white roots embedded in rich elephant dung—one leaf light green, the other pale pink. They are the innocence of new life, fragile yet tenacious in their pursuit of light. Beauty is in the scent of the Jeffrey pine tree I inhale when I lean against its rough trunk, pressing my nose between the bark's ridges. Caramel. Vanilla. Butterscotch. Beauty is the bright lipstick-red *Aframomum* fruit skin pulled back to reveal white feathery pulp around dark seeds that taste like tart lemonade. Beauty is in the cry of the hawk, carving the sky until the sun lights its tail feathers a fiery russet red. Awe at the astounding ability of nature to create and recreate beauty is one of the many roots of passion fundamental to the desire to protect nature.

When I speak with Purnima Devi Barman about the roots of her passion to protect nature, she tells me about growing up on the banks of the Brahmaputra River in India. Purnima recounts how her father worked in the army and was transferred from one place to another. It was hard for her parents to move so often with a small child, so when she was five years old, her grandmother offered to take care of her. After her parents left, Purnima was inconsolable. She explains, "Our home was made of only clay, and we didn't have food in our house. It was very difficult for my grandma, but she found a solution for me." Every evening, Purnima's grandmother took her to the paddy fields. "We would watch the birds fly and she would sing to me," Purnima says. "I still remember the songs. We saw many birds like vultures and storks. She told me the names of them, but they were the local names. Do you believe my grandma was married at nine years old? Just imagine, she was nine years old. She didn't have a chance to go to school. She didn't know how to write, but she had this deep connection with nature."

From that moment on, Purnima's grandmother took her to the fields to see the birds every evening. Purnima says, "Slowly, I became stable again." In her adult years, she became a protector of the greater adjutant stork—one of the birds she once watched on the banks of the Brahmaputra River. Purnima founded the "Hargila Army" of ten thousand women across Assam who work to protect the stork. (Hargila is the Assamese name for this stork, which means "bone swallower.") Because of Purnima's efforts, the number of storks in the area where she works has increased to become the largest colony of the greater adjutant stork in the world.

For Reshu Bashyal too, research fellow at Greenhood Nepal, her aspirations to cultivate a career in the environment stemmed from a childhood growing up in the Palpa district in rural Nepal. She says, "It was not as happy as it was for others, but still, we managed to stay optimistic. When I was quite small, I was responsible for collecting water, fodder, and everything like other village kids do. I had to collect water from a spring which was around a one-hour walk, one way; normally we would be with other girls and women. As a result, I had a special connection with the water. I also had a special curiosity about plants."

In a rural area with limited health facilities like the one where Reshu grew up, people seldom visited doctors; most of the treatment would occur at home using locally available plants. Reshu's grandfather understood a lot about plants and their medicinal properties, and people came to him to treat their ailments. Reshu recalls an incident one day when both her grandparents fell sick, and they asked Reshu to find a specific plant for them by visiting a shaman in another village. Seeing the miracles that plants could perform instilled in her a deep wonder and respect for nature. As she grew up, she realized some of these plant species were harvested on such a massive scale that some might go extinct, which meant the potential loss of the plants' ability to treat and heal illnesses.

Today, Reshu works to understand how the demand for traditional medicines affects the trade in wild species in Nepal. One of her passions is orchids. She tells me about a type of orchid that mimics a female bee, with soft pink petals that imitate its wings and stripes that traverse the velvety lip of the flower in imitation of the bee's body. This, and the flower's mimicry of two dark eyes and a yellow powdery antenna, attracts male bees. Reshu also tells me of other delicate orchids that take the shape of a monkey, a swan, and a ballerina, and how they are used in traditional and Chinese medicine, cosmetics, ice cream, and even as an aphrodisiac.

Other women also spoke about how their passion for nature was rooted in childhood, but that the disappearance of nature led them towards a career in protecting it. Farwiza Farhan, named one of TIME100 Next in 2022, grew up in Banda town in Indonesia. She says:

> In my childhood, I spent a lot of time outside, climbing trees, and swimming in the river. When I was in high school, there was a civil war happening in this area. Somebody threw a bomb at our house. It was a small, homemade bomb that did not explode, but it burnt a patch of grass on our lawn, and my mom decided that it was no longer sensible to live there, so we moved to Jakarta. There I came to the realization that nature is not always accessible—that swimming in the rivers is something that you cannot do every day, because rivers

> there are heavily polluted. You can't just run around outside into a patch of forest behind your house or climb trees because there are not many of them. All these things suddenly became quite foreign.

For Caroline Achieng' Ouko, senior research scientist and training coordinator at the Centre for Training and Integrated Research for ASAL (Arid and Semi-Arid Land) Development in Kenya and co-founder of the Kenya chapter of the Women in Nature Network, it was how nature changed and in some cases disappeared that launched her on a career path to protect it. Caroline says, "What really connected me to nature and the work that I do today is the river we used to cross when going to school in Nairobi. This river used to have a lot of water and sounds. What amazed me were the two seasons. When it was dry, the louder cries were different than when it was raining, when there were all manners of noises. But as we grew up, they stopped. There was no life in the river anymore. I wanted to know what happened. Why did the river die? And this experience is one of the key things which propelled me into conservation."

For many women, however, a close relationship with nature is not always one of choice. Instead, sometimes, women are thrown into relationships with nature by circumstances of dependence and poverty.

On April 22, 1991, a storm was gathering force over the Bay of Bengal in the northeastern Indian Ocean. By April 29 it was a "super cyclonic storm"—the highest category of cyclone possible. With winds of up to 240 kilometers per hour, it made landfall at high tide along the low-lying coastal area of Bangladesh where millions of people lived. Relentless winds pounded the shore for twelve hours. The sea surged, flooding the land and killing about 140,000 people. Of the dead, an astonishing 91 percent were women.[2]

In Nepal, during the 1993 floods, women were almost one and a half times more likely to die than men.[3] In the 1995 earthquake in Kobe, Japan, more women died than men.[4] In 2004, a devastating tsunami crashed into the coastal areas of Southeast Asia, South Asia, and East Africa, killing more than 230,000 people across twelve countries. According to one account, "When the survivors of Lampu'uk had picked themselves up out of the mud of the tsunami, several appalling facts became clear. The first was that their town no longer existed. The second was that four out of five of its former inhabitants were dead. But it took a while to understand the strangest thing of all: that among those who made it to higher ground, or who kept their heads above the surging waters, so few were women."[5] In total, 70 percent of those who died because of the 2004 tsunami were female.

When Cyclone Nargis hit Myanmar in 2008, 61 percent of fatalities were women.[6] In New Orleans, most victims trapped because of Hurricane Katrina were women and children.[7] During the 2003 heat wave in Europe, nearly two-thirds of those who died were women.[8] In the 2009 tsunami that hit Samoa and Tonga, about 70 percent of the adults who died were female.[9] During the 2010 heat wave in Ahmedabad, India, more women than men died.[10] During 2013's Typhoon Haiyan in the Philippines, about 50 percent more women died than men. Anecdotal evidence from the 2014 Solomon Islands flash floods suggests that 90 percent of those who died were women and children.[11]

It is overwhelmingly clear from the data that climate change does not affect everyone equally. There are also instances when men are affected more, and we will explore these later. However, in most cases women suffer disproportionately from our warming planet. The terrifying fact is that things are going to get worse.

Today, the amount of $CO_2$ in the atmosphere is higher than at any time in at least 2 million years. The planet is now about 1.1°C hotter than at the start of the Industrial Revolution, and since 1900, the mean global sea level has risen faster than in any other century since at least the last three thousand years.[12] The number of record-high temperatures globally has risen, and the number and severity of climate-related extreme events involving storms, heat waves, droughts, floods, and fires has quadrupled in the last two decades.[13] Climate change is affecting the Earth's crust, too, increasing the number of earthquakes and volcanic eruptions, which in turn causes more landslides and tsunamis.[14]

It is not just in the moment of crisis that women are unequally affected. In the aftermath of droughts, floods, tsunamis, fires, and disease outbreaks, women continue to bear the brunt of the impacts. Temporary camps providing shelter to those displaced by natural hazards are often crowded and lacking security. Women are asked for sexual favors for their share of food distributions, and women and girls often suffer increased physical and sexual violence.[15] An increase in gender-based violence was documented in Canterbury, New Zealand, after the 2010 and 2011 earthquakes,[16] in Haiti after the 2010 earthquake,[17] in the Philippines after 2013's Typhoon Haiyan,[18] in Pakistan during the 2011 floods,[19] in Japan after the 2011 tsunami and Fukushima disaster,[20] and in Sri Lanka after the 2004 tsunami.[21] The United Nations estimated that 3.2 million women and children were at risk of rape, sexual exploitation, violence, and trafficking after the 2015 earthquakes in Nepal.[22] Domestic conflict significantly increased in Louisiana after the 2010 BP oil spill.[23]

The chaos following natural hazards can also result in unwanted pregnancies. For example, between 2020 and 2021, women interviewed in Mozambique and Bangladesh reported loss of access to birth control after floods, either because medicines got washed away, they didn't pack them during evacuations, or they had difficulty accessing reproductive health clinics and pharmacies because the buildings were damaged or closed, or transportation to get to the clinics was inaccessible.[24]

Climate change is also impacting the health of pregnant mothers and babies still in the womb. A warmer climate and increased flooding are expanding the prevalence of mosquitos and vector-borne illnesses such as malaria, dengue fever, and Zika virus. These viruses can put both mother and baby at risk.[25] For example, Zika virus infects the baby in utero and can result in damage to a baby's eyes and brain. Many babies infected with Zika virus are born with microcephaly—a birth defect where a baby's head is smaller than that of an average healthy baby. Extreme heat during pregnancy can increase the risk of early labor and stillbirths.[26]

Everyone suffers psychological trauma during and after natural hazards, but studies are uncovering important differences in the way that women and men are affected. Women have higher risk than men of suffering from certain anxiety and emotional distress.[27] For example, after flooding, women in Bangladesh, India, and Nepal demonstrated more psychosocial effects than men as a result of having to care for other family members in addition to being under stress themselves. Also, women often lose their support networks when they must relocate to safer locations.[28] Women are also in greater danger of starvation,[29] which renders women increasingly vulnerable to future events as well.[30]

Devastating climate-driven events often receive more attention, but it is also the drying up of rivers, the disappearance of forests, and the death of wildlife that is taking its toll unequally on women. Women and girls are spending longer searching for and fetching water and fuelwood, stealing time away from girls' education and women's paid work. In addition, when women and girls have to walk farther, or walk at dawn, dusk, or night, they are exposed to further risks.[31] For example, one study documented that not only are women having to walk farther to find fuelwood in Kenya, but when the wood debris around the edges of forests is exhausted, they are forced to enter forests, where they have reported being attacked by elephants.[32]

Aside from risks to safety, there is a physical toll placed upon women. In many countries women frequently carry wood weighing up to seventy-five pounds. When fuelwood gets scarcer, women must walk farther with these heavy loads on their heads; this physical burden has been linked to spine damage as well as complications during pregnancy and even maternal mortality.[33] Walking alone farther from home also increases the chances of being sexually harassed or assaulted.[34] In Darfur, for example, women and girls have been attacked while gathering water.[35]

When food is short, and there is little income because of fewer resources, girls and women sometimes resort to sex work to earn money.[36] With less food in the house, families are increasingly making the decision to marry their daughters early so that there is one less mouth to feed. As a result, the number of child brides is increasing, as is gender-based violence.[37]

Gender-based violence and violence against women and girls has reached a crisis level, with one-third of women reporting to have experienced gender-based violence in their lifetime. The World Bank has called it a global pandemic.[38] Gender-based violence can take many forms and is linked to the environment in many complex ways—both directly and indirectly. It often occurs as an attempt to control access, such as when sexual favors are demanded for land or resources.[39] Gender-based violence is often intertwined with environmental crimes, such as illegal logging, mining, and fishing. In extractive and large-scale infrastructure projects, gender-based violence has been used as a form of power and intimidation. It has been used by park rangers against local communities.[40] Environmentally related gender-based violence is especially hard on Indigenous women and girls, who are often at the forefront of defending their community's resources and land from extractive industries, and it has been used as a "tactic to silence their activism and work."[41]

These intersectional and unequal impacts of the destruction of nature result in reverberating repercussions that last a lifetime, so much so that women's overall life expectancy can decrease as a result.[42]

Pandemics, too, have disproportionate impacts on women, COVID-19 being just the most recent example. Viruses that cause diseases such as Ebola, HIV, and COVID-19 come from nature, and they are increasing in frequency as a result of natural resource exploitation and climate change. Pandemics are often the result of viruses crossing over from nonhuman animals to humans. In *Spillover: Animal Infections and the Next Human Pandemic,* David Quammen describes, in fascinating and sometimes gruesome detail, how the HIV virus "spilled over" from one Cameroonian chimpanzee to humans, how the Ebola virus "spilled over" from a bat to a child in Guinea, West Africa, and how so many other diseases, such as the Marburg virus, SARS, avian influenza, and Lyme disease, all are a result of these spillover events in which viruses transfer from their animal or insect host to humans.[43] As humans encroach on and destroy the habitats of wild species, as human-wildlife interactions increase, and as species that historically would have never come into contact with humans are forced into proximity with us in wildlife markets, the pet trade, and through globalized trade of animal parts, the frequency of pandemics is going to increase.

Pandemics often initially result in greater mortality for men. However, the aftershocks of pandemics have shattering and long-term effects on women that cannot be ignored. For example, although the Ebola outbreak in West Africa initially killed more men than women, the indirect health, social, and economic consequences for women were profound.[44] Similarly, although male mortalities (55 percent of all deaths) from COVID-19 were higher than for women (45 percent of all deaths), the longer-term impacts on women have been devastating.[45] Job and livelihood loss, domestic violence, unwanted pregnancies, school drop-out rates, and trafficking of girls and women all increased.[46]

As with climate change, COVID-19 increased teen pregnancy rates in some countries. Early teen pregnancies put girls at tremendous health risk. For a girl to carry a baby and give birth before her body is ready can be life-threatening. Teen pregnancy rates increased during the first year of the pandemic in both South Africa and Ghana, but not in Brazil.[47] In Kenya, girls experienced two times the risk of becoming pregnant before completing secondary school because of COVID "containment measures."[48] Faith Milkah Muniale, programmes manager at the Tropical Biology Association, talked with me about the impacts of lockdown on girls in Kenya:

> The pandemic—it's such a disaster, it took us ten years back with the progress that we had made. In Kenya, women really suffered during lockdown because of gender-based violence. A lot of women also suffered harassment, even small girls. We recorded a very high statistic in Kenya of teenage girls who got pregnant in lockdown because they were not going to school, and they had been raped by older men. In one district, there were sixty girls who were supposed to be in school, but the school was closed and in the course of all that, many got raped. They got pregnant. They are mothers at fifteen and sixteen years old. Now that we are seemingly getting a solution for the virus, it's a call for all of us to go to another layer and start addressing the ripple effects of the pandemic. For example, now so many girls have children who they are not able to raise, and this is just

creating another generation of potentially unhealthy people.

HIV infection rates, although once highest in men, are now highest among women in many countries. For example, in South Africa, a 2015 study revealed that the HIV rate among females between fifteen and twenty-four years was over four times higher than for males of the same age.[49]

The impacts of pandemics are not equal by race either. Analyses by the Centers for Disease Control and Prevention reveal that both Native American or Alaska Native individuals were over two times as likely to be hospitalized or die from COVID as white, non-Hispanic individuals. Black or African Americans were over two times as likely to be hospitalized for COVID and 1.6 times as likely to die of COVID as white, non-Hispanic persons.[50]

It is not only women but people who do not neatly fit into gender binaries who face heightened discrimination and suffering because of climate change. For example, in the United Kingdom, almost one-quarter of homeless people are LGBTQ+, and homelessness increases vulnerability to the impacts of storms, floods, and heat waves. In the aftermath of natural hazards, LGBTQ+ people can have more trouble crossing borders, and they are sometimes excluded from emergency shelters or refugee camps. Many countries do not recognize LGBTQ+ rights, which can leave people extremely vulnerable after a natural hazard. After Hurricane Katrina in New Orleans, people in same-sex marriages whose partner had died were not recognized for property inheritance or insurance claims.[51] LGBTQ+ people are sometimes even blamed for the occurrence of natural hazards. For example, Maurice Mills, then mayor of the Irish town of Ballymena, said Hurricane Katrina was "God's wrath" against the LGBTQ+ community in New Orleans.[52]

Often we talk about the extinction of a species, a devastating flood or fire or the spread of disease across human and wildlife populations, as an isolated event, yet climate, biodiversity, and health are all interrelated. I liken their interaction to that of the sides of a polygon. As one side is altered, it changes the other sides and transforms the overall shape of the whole. What we are our experiencing today has been called a planetary "polycrisis."[53] As any one of the crises expands, our world as we know it transforms as a result. All these crises are interconnected.

The inequalities of our planetary polycrisis are intersecting too. The number of female deaths due to climate events are directly linked to women's economic standing, but they are also linked to their social rights.[54] Where women and men have equal rights, natural hazards have similar impacts by gender. In other words, climate change has a greater overall impact on those people who are already marginalized, with compounding and intersecting impacts.

The inequities are clear, yet the attention to more impacted groups is not proportional. For example, a 2016 study led by Craig Leisher, a director of monitoring and evaluation at the conservation organization The Nature Conservancy, screened over eleven thousand records about conservation projects in Scopus, CAB abstracts, AGRIS, AGRICOLA, Google Scholar, and Google, as well as websites of international conservation organizations, and found only seventeen articles that fit their inclusion criteria of connecting "women" or "gender" to conservation outcomes.[55] Although the impacts of climate change are much harsher in the Global South, a 2021 article in *Nature Climate Change* found that there has been less research investigating climate change impacts in the Global South than in the Global North.[56]

Why are we not paying more attention to these issues of inequality? The short answer is that we are not paying attention for the same reasons that these inequalities and vulnerabilities exist in the first place.

# 2

# VULNERABILITY

> Climate-induced disasters don't blast away existing social relationships and inequalities, as if out of the rubble comes a chance to reset. They do the opposite: what was unequal before becomes even more unequal after.
>
> —*Anne Karpf*[1]

Damien Mander is the founder of the Akashinga, an all-female ranger patrol in Zimbabwe. In his twenties, Damien was a clearance diver (a diver who specializes in clearing explosives under water) in the Royal Australian Navy and a special operations sniper in Iraq. After three years in Iraq, he came to Africa at the age of twenty-nine. He used his skills to train an all-male anti-poaching patrol in Southern Africa. To select the team, he put the men through a rigorous test, subjecting them to three days of hunger, thirst, sleep deprivation, and intense physical exertion. "I did this selection course with 193 men in 2011," he says. "At the end of day one, we had three men left."

In 2017, Damien formed an all-female anti-poaching team in Zimbabwe. A total of eighty-seven women came to the first recruitment screening. Most were single mothers and survivors of sexual assault, domestic violence, or AIDS. Damien and his team quickly realized that the process they had used for the male rangers—to weed out the strongest—was not going to work for selecting the all-female team. He tells me, "We had a team of instructors who had only ever worked with men before. Very quickly, we became aware that we were working with something very different. Absolute, raw grit. What we put them through in three days was nothing compared to what these women had gone through in their life. Only three women dropped off voluntarily during those three days."

The Zambezi Valley in Zimbabwe is a vast floodplain for the Zambezi River—the fourth largest river in Africa and home to crocodiles, zebra, lions, hippos, buffalo, and elephant. Across Africa, poaching for meat and ivory is the most serious threat to the survival of elephants. Since it was launched, the all-female Akashinga team has made extraordinary progress in decreasing the level of poaching in the Lower Zambezi Valley. Within the first two and a half years of operation, the team made 191 arrests, contributing to an overall 80 percent decline in elephant poaching in the area. The program has now grown to 240 women enlisted, and it encompasses five parks covering 4,000 square kilometers, with a goal to train 1,000 women as rangers in twenty parks by 2025.[2]

I spoke with Nyaradzo Hoto, an Akashinga sergeant, a One Earth "Climate Hero," and winner of the 2022 IUCN International Ranger Award. I wanted to hear her perspective on the training and selection process she underwent to be on the team. Nyaradzo says that, at the beginning, the biggest challenges were not in fact the physical ones, but the attitudes of people toward the women. She tells me, "We were so excited and interested to be part of the team. At that time, I can say that the only challenge that we faced was the misconception of the community. They didn't believe that women could be rangers. The men were mocking us and saying bad things about us. And they said that we were taking over their duties. But now people are understanding why we are doing this work. Through our success, through the arrests we made, and through the projects we are doing around the concession, the population of wildlife has increased."

Rachel Ashegbofe Ikemeh is another example of raw grit. Rachel is founder and director of the South West/Niger Delta Forest Project and winner of the Whitley Award in 2020. Rachel didn't start out as a conservationist, but after she received a mandatory placement in the National Youth Service Corps in southwest Nigeria and saw a seemingly endless number of large trees being brought to a nearby sawmill, she became curious about the environmental impact of the wood business on tree populations at logging sites. After her placement, Rachel took a job with a national nongovernmental organization (NGO) that at the time worked in Gashaka-Gumti National Park, among other natural areas. Rachel was fascinated with the lions in the national park, because very little was then known about them. She proposed to her supervisor that she lead a survey to find out how many lions lived in the park. He agreed. Without much training, she set off for thirty days with a team of men, sometimes walking up to twenty kilometers a day looking for evidence of lions.

Rachel described what it was like being a young, inexperienced woman doing fieldwork for the first time: "I was a skinny, skinny girl. I was very young. The men thought, 'This is going to be fun!' So they joined me, thinking that they would just watch it play out and see how long the survey lasts, believing that they would be back home again soon. I felt like they had their eyes on me, looking for a weakness that they could capitalize on. I was in a situation where I was the most vulnerable in my entire life."

But Rachel was undeterred:

> It was two weeks into being in the field that it dawned on those guys that this wasn't going to be a short trip. I was going to see it through to the end, and there was no stopping. It was about this time they started getting curious. They thought what I was doing was impossible. They started talking amongst themselves, wondering if I was doing drugs. They thought it might be something in the cup of tea which I would take every morning, but it was just Lipton tea in hot water, no sugar, no cream. And that's all. Then we would walk sometimes thirteen to twenty kilometers in a day in the savanna. In the evening, I would eat two packs of noodles. And that was my daily routine for the four weeks we spent in the heart of this vast wilderness. One morning, I was sipping my tea and they wanted to test it, because, like I said, they thought I was putting something else in it that somehow was boosting my energy levels. One of the guys asked for a sip. I said fine, and he sipped it. Of course, there was nothing in the tea, so they said amongst themselves that maybe I took something when I was having my bath by the stream because that was the only time they couldn't see me. In the end they just concluded that and comforted themselves with that assumption.

At one point during the survey, Rachel suffered from an illness that caused temporary blindness for a few hours and affected her knee joints for some days. Determined not to stop the survey, she used a cane and continued walking many kilometers each day. Rachel and her team conducted the country's first scientific lion survey. She says, "At the end of the survey, the men had a renewed respect for me and the work I was doing. They all looked at me differently from that time on." Since then, Rachel has braved many storms during the course of her career that she believes have made her resilient and helped shape who she has become as a conservationist. Today, the NGO she leads has worked to create, protect, and manage two protected areas in Nigeria, and she has spearheaded efforts to bring back critically endangered red colobus monkeys from the edge of extinction.

Nyaradzo and Rachel are but two of the extraordinarily strong women I have spoken with in recent years. Never once in my conversations did the word "vulnerable" come to mind. "Feisty," "bold," "determined," and "courageous"—but not vulnerable.

There is essentially nothing innate about women that makes us more vulnerable than men to the climate and biodiversity crises. Very few physical differences are responsible for the unequal ways in which we are impacted by climate change. In fact, people with XX chromosomes, in many ways, are better built for survival. We live longer and are more likely to fight cancer, sepsis, infections, and trauma.[3] We have less susceptibility to obesity, heart disease, stroke, and diabetes.[4] We have stronger immune systems and are less likely to suffer from certain genetically inherited diseases, because one X chromosome can mask mutations on the other X chromosome. That certainly doesn't make us sound like the weaker sex.

We are not physically more vulnerable. Our bodies are capable of spectacular things. During pregnancy, our blood volume doubles. From one fertilized egg we can grow a baby with 26 billion cells within our wombs, and we can push that baby with a head width of 13.75 centimeters out of our bodies and into the world. No, we are not fragile. Women's fragility is a myth, and belief in this myth can keep us from fulfilling our full potential.

The other myth is that our brains are different. No matter how hard scientists try to find differences in our brains, decades of research have led them to the conclusion that the brains of male and females are not sexually dimorphic.[5] Our brains are organs just like hearts or kidneys. Our brains do not have a sex. In fact, there is more variation in brains between women, and between men, than there is between women and men. Our brains do not make us more vulnerable either.

"Vulnerability" as defined by climate scientists, however, means something different from "weakness." In climate science, "vulnerability" is used to describe susceptibility to be harmed by, or to have to cope with or adapt to, the impacts of climate change.[6] Under this definition, women are indeed more vulnerable to the climate and biodiversity crises, although, to be clear, this vulnerability does not come from our sex. Instead, women's vulnerability comes from the highly gendered division of labor in societies, and the roles, rules, limits, controls, and devaluing of women that result from social constructs.

Vulnerability comes from the fact that women bear the brunt of housework and childcare. Vulnerability comes from society limiting women's access to education, birth control, money, knowledge, skills, and jobs. It comes from society controlling what women wear, where they can go, and with whom they associate. Vulnerability comes from messaging, since birth, that you are valued less. Critically, vulnerability comes from poverty.

Women make up the majority of the world's rural poor.[7] Poverty increases women's dependence on natural resources, which also increases their vulnerability to the climate and biodiversity crises. In many countries, people are dependent on stream water for drinking, cooking, washing, and irrigation. People are dependent on wood from forests for the bare necessities of heat, light, and cooking. Wild animals, fish, plants, fungi, and tubers provide food. Fruits, bark, and leaves provide medicine. As these resources decline, women who depend on them are deeply affected by their loss, in every aspect of their lives. Poverty locks women into an ever-increasing state of vulnerability.

Poverty is not just a matter of income. Poverty is complex, multidimensional, and ever-changing, and it hurts people in different ways. It takes the form of hunger, malnutrition, and limited access to education. It results in social discrimination, exclusion, and the removal of women from decision-making. It steals agency and dignity.

Poverty looks different among women. There are variations in how women are affected depending on factors like race and geographical location. Because poverty manifests itself in so many ways, it is something very difficult to measure. The way that poverty has historically been measured has made it challenging to understand its impact specifically on women.

Traditionally, poverty measures were based on surveys of household spending on things like food, housing, and utilities, which was then compared with household income. Using this data, the World Bank identified for each country a "national poverty line," and household income and spending were compared with this threshold..If the household was below the threshold, that household was categorized as "living in poverty." "Extreme poverty" is when a household earns less than $2.15 per day, a threshold calculated by measuring the national poverty lines across fifteen of the world's poorest countries.[8]

However, because not everyone within a household has the same access to resources it is very difficult to discern the poverty experienced specifically by women. Lumping together everyone in a household disguises individuals' differing experiences. Women sometimes receive less food than others in a household, for example, and daughters in a family may not have as much access to education as sons. Women may not have the same access to health care or money as their husbands.[9]

Measures of poverty have been refined, taking into account

poverty's complexity and many more of its dimensions. Two such examples are the Societal Poverty Line (SPL) and the Multidimensional Poverty Measure (MPM). The SPL accounts for the changes in the way that poverty is experienced as nations become increasingly rich. Calculations of MPM incorporate measures of well-being, such as education and "access to basic infrastructure," extending beyond monetary poverty.[10]

In a 2021 study of poverty, the World Bank tried to tease apart differing experiences of members within a household by disaggregating data by gender and age.[11] This analysis revealed that, globally, girls and boys start life experiencing poverty at similar rates, but by the teenage years girls are already experiencing greater poverty than boys in most regions of the world. The greatest difference in poverty levels between women and men occurs in the age range between twenty-five and thirty-four, and the gap is widest in South Asia and sub-Saharan Africa. Women in this age group are most likely to be poorer, primarily because this is a period in their lives when they are most likely to have children, and "living with children" is a factor associated with higher poverty for all adults. Overall, women are more likely than men to live in households with children.

Lone-mother parenting has been increasing in all developing regions since the mid-1990s. In sub-Saharan Africa, Latin America, and the Caribbean, lone-mother households make up about 10 percent of all households. For comparison, the percentage of lone-father households is between 1 and 2 percent.

Reshu Bashyal, the biologist studying the plant trade in Nepal, notices this trend in the communities in which she works: "Men are normally the ones who go abroad to earn money, and rural women are the ones who are left behind. They are supposed to manage their household, do all the daily chores, feed their livestock, and collect fodder from the forest. If they have businesses or small enterprises, they have to manage all those things. As a result, women are impacted when anything goes wrong. For example, if there is a flood or landslide, women are the ones who have to look after the children and elderly because the men have gone to earn money. Women are the ones who are fighting to manage their families."

Willemien Le Roux, founder, board member, and director of Pabalelo Trust, also speaks with me about the increasing isolation of women living in rural areas in Botswana as a result of climate change. Willemien was born in South Africa and moved to Botswana at the age of fourteen to live with her missionary parents. She later left home to go to college, and worked as a journalist in Johannesburg, but after twelve years returned to Botswana with her pastor husband. When her husband passed away, she stayed on living and working with the San communities. I speak with Willemien on Zoom, as the sun is setting behind her in her home near the Okavango. She explains:

> Women are the main food providers, especially in an era where hunting, which was a traditional role of food provisioning for men, has been taken away. Hunting and protection of your family by spear or bow and arrow are not allowed anymore, so the men go to find employment and income and the women are left entirely with the role of food provision. With climate change, it has become such a hard life because the men are no longer there to help with cultivation of the fields. They would have been around to chase away the hippos and elephants but now they are not. Young people also move away from their homes because there's no food, so that puts women into an even more vulnerable position. In the past, they would have had teenagers around who could help them do physical work, fetch water, and get firewood.

She adds that with climate change, "the weather has become so erratic, there is no surety that you will have a season for planting every year, and the rain comes later and later. And if it comes, it often comes in huge masses or it becomes too widely scattered. The women are the ones who bear the brunt. Now they sit with the hungry children."

Whether it is because women are the sole parent in the household, or because of societal expectations, women are, almost universally, the main caregivers for the young, elderly, and sick, and women take care of the home. Being a primary caregiver and home caretaker contributes to a woman's poverty in many intersecting ways. First, most of this kind of labor is both invisible and unpaid. Although there is a lot of variation within and between regions, women worldwide on average do three times as much unpaid care and domestic work as men.[12] The gender gap in the amount of unpaid work is greatest in Northern Africa and Western Asia. In India, for example, men do an average of thirty-six minutes of unpaid care work per day, compared with six hours for women. In Ghana, women generally do between two-thirds and three-quarters of the household chores and two times the amount of childcare than men.[13] As a result, regardless of marital status, age, or income, on average, women in Ghana work thirteen hours per day, and 60 percent of that work is unpaid. In fact, the amount of time women and girls spend doing unpaid work globally each day totals a stunning 25 billion hours.[14] The unpaid care work of women is worth about $10 trillion per year.[15] While some people take issue with putting a monetary value on unpaid care work, others feel it is a useful exercise to shine a spotlight on the hidden contributions of women.

Another reason that care work contributes to women's poverty is that it absorbs time that could be spent doing paid work. Globally, there are 700 million more men in paid work than women.[16] When women do have paid work, on average and globally, they continue to do most of the household chores. In fact, a 2019 study reported that 43 percent of women who were the primary income earners for their household continued to do most or all the household work.[17]

When women are expected to do most of the household chores and childcare, they often must resort to part-time instead of full-time work. This is especially true for women between the ages of twenty-five and forty-four who are more likely to have young children. Partly because women are working part-time, about 75 percent of their jobs are considered part of the "informal economy," which means the work is not recorded, monitored, or taxed. This proportion varies by world region. In South Asia, for example, over 80 percent of women's work is informal. In sub-Saharan Africa, that proportion is 74 percent, and in Latin America and the Caribbean, it is 54 percent.[18] One of the downsides to working in the informal economy is that jobs are unlikely to have work contracts or social benefits, and as a result, women often have fewer legal rights and security if, for example, they get ill or are temporarily unable to work.

Even when women hold the same job as men, they earn on average 23 percent less.[19] The pay imbalance does not reflect merit or qualifications either. In the United States, for example, in 2021 women who had earned a master's degree still earned about 38 percent less than men with the same qualifications.[20] In addition, many studies have now confirmed two phenomena known as the "fatherhood wage premium," where men who are fathers earn more money, and the "motherhood penalty," where women who are mothers earn less, as a result of biases toward both groups.[21] While mothers are considered less able to perform their work duties, and less committed to their work once they have children, men are viewed as more committed and productive once they become fathers.

Women sometimes have less money because men claim the responsibilities for holding and spending it. Approximately one in three married women in the Global South have no control over household spending on major purchases, and one in ten are not even consulted on how their own earnings are spent.[22] In addition, women can be more vulnerable to poverty because they have greater difficulty accessing credit or loans that provide buffers during difficult times. Women also face barriers to saving their money in banks and other financial institutions.[23] Having savings, of course, helps to build financial resilience to ride out difficult economic circumstances caused by job loss, property loss, health problems, or an extreme weather event, such as a flood or fire, that has affected their livelihood or destroyed their home and other assets.

Unfortunately, women often do not have the same access to technology that could help them make more money. Although women contribute 60 to 80 percent of the labor for food production in many low- and middle-income countries, they do not benefit as much as men from technologies that could improve their yields.[24] Women are frequently excluded from participating in training that could decrease their vulnerability as well. For example, in government training courses in aquaculture for fishing communities in Malaysia, there were only 18 women out of 952 participants between 1996 and 2001. One study revealed that lack of literacy and women's inability to leave their household chores were the main reasons why so few attended the course.[25]

Poverty, of course, is not just about money. It is also about not having other assets as well—primarily land. In many countries, women are legally or culturally barred from owning land. In fact, only about 15 percent of landowners worldwide are women, although this proportion varies widely between countries.[26] For example, in Saudi Arabia, less than 1 percent of the land is owned by women, compared to almost 35 percent in Botswana. A study in India determined that about 10 percent of plots of land across nine states are owned by women, 2 percent are owned jointly, and 88 percent are owned by men.[27]

When women do not have access to money or land, trading goods becomes a critical means of getting food. As resources get scarcer, however, women in some circumstances are resorting to transactional sex to feed their families. This is most common across parts of Melanesia, sub-Saharan Africa, and Asia.[28] Under such conditions, women are often unable to negotiate safe sex. Some fishing communities across the world have now become "hotspots" for HIV/AIDS, especially in Asia, Africa, and Latin America, where collectively over 95 percent of the world's fisherfolk live. HIV rates in fishing communities in the Global South are four to fourteen times greater than the national averages and are increasing more for women than for men in these areas.[29] For example, in the Pwani region of Tanzania, women are three times more likely than men to be infected by HIV.[30]

These devastating statistics demonstrate how women are unequally impacted by biodiversity loss and climate change, largely because of poverty, but it is critical that women not be portrayed as victims without agency. Media, humanitarian, conservation, and refugee organizations often use images of women struggling in the Global South to help in fundraising efforts. However, such images of women solely as victims portray them as weak and perpetuate gender and racial stereotypes. Without acknowledging the root causes of climate change and social inequalities, such images circumvent the culpability of the Global North in creating climate change and poverty, steal agency from women, and feed stereotypes of women as less capable. Anne Karpf, author of *How Women Can Save the Planet*, writes, "When we obscure the enduring historical links between the wealth of the global North and the poverty of the global South, the gendered inequalities deepened by climate change come to seem like geographical or biological facts of life that can be improved after the fact by charity or philanthropy."[31]

The news is not all bad either. According to the World Bank, poverty has been declining for thirty years, and global rates of extreme poverty have been cut by more than half.[32] It is not a time for complacency, however. COVID-19 was a significant setback to addressing poverty, and at current rates, we are unlikely to meet the UN's Sustainable Development Goal of ending extreme poverty by 2030.[33]

Poverty increases women's vulnerability to climate change and the biodiversity crises, but if we look even deeper into the root causes of women's poverty, the situation is more complex than this fact alone suggests. Swirled into the mix of poverty and reliance on natural resources are the expectations, roles, and responsibilities piled upon women worldwide and the cultural restrictions that lock them into never-ending cycles of poverty—thus increasing vulnerability to climate change and resource depletion even further.

# 3

# WOMEN'S WORK

> The work women are made to do creates sensitivities and empathy. As a result, there is a moral knowledge present for women, not because of being in a female body, but because of what female bodies are made to do.
>
> —*Chris Cuomo*[1]

Women are not a homogeneous group. In some parts of the world and for some cultures, women are not forced into gendered roles. However, in many parts of the world certain tasks, such as the responsibility of caring for the young, sick, and old and being the primary collector of water and fuelwood, fall primarily to women, which can increase their vulnerability to a changing environment. These gendered roles can make a difference in how women respond to emergencies and extreme weather events.

During a natural hazard that requires rapid evacuation, such as a fire, flood, or tsunami, women as the primary caregivers are often prevented from fleeing quickly, easily, or at all.[2] When sick or elderly people can't be moved, women often remain to care for them. In New Orleans, women made up 80 percent of those left behind to face Hurricane Katrina, even after the orders to evacuate were given (although women represented about half of the city's population).[3]

Many studies have documented that when food is short, women will almost universally prioritize feeding their children first and consume fewer calories themselves, probably as a result of their being the primary caregivers for children.[4] This happens either by choice or because of cultural hierarchies regarding food that require women to eat last. For example, in a 2021 survey by CARE International (Cooperative for Assistance and Relief Everywhere) in Lebanon, 85 percent of women reported eating smaller portions, compared with only 57 percent of men, during food shortages. A similar study of how households in Afghanistan were responding to drought found that "the newest female in the household (generally a daughter-in-law) is likely to be more food insecure, as she has the socio-cultural role of ensuring all other senior members of the household are cared for before she can care for herself."[5] During the 1994–95 droughts in Zimbabwe, the body mass of women (but not men) decreased.[6] Malnutrition has both short- and long-term consequences for women's health.[7] When women have poorer nutrition, they are more susceptible to diseases, less likely to be able to escape from future natural and anthropogenic hazards or survive further food shortages.[8]

Women's responsibilities for caring for the sick also make them more vulnerable to contracting disease during pandemics through experiencing direct contact, being physically near patients, and cleaning patients' bedding and other items According to the U.S. Bureau of Labor Statistics, 87 percent of registered nurses in the United States are women.[9] While the COVID-19 pandemic resulted in more deaths of men than women as a direct result of the illness, there were many reverberating impacts on women partly because of the types of jobs that women hold. By May 2020, more than half of the jobs lost worldwide belonged to women, even though women's jobs were only 39 percent of total employment. As a result of children being out of school, unpaid care work also increased for women.[10]

Seeking and carrying water and fuelwood are also traditionally considered women's jobs in many countries. In eight out of ten homes without running water, girls are responsible for fetching water. UNICEF estimates that girls and women worldwide collectively spend a total of about 200 million hours a day gathering water.[11] Women in sub-Saharan Africa collectively spend 16 million hours a day collecting water, while men spend only 6 million hours.[12] A report based on data from eighteen African countries found that women were five times more likely than men to collect drinking water for the household.[13]

As the planet gets hotter and water becomes scarcer, women and girls are walking farther to find water. For example, in recent years in Guatemala, women are spending up to eight hours a day getting water, rising at two in the morning to have enough time to fetch water during dry spells.[14] During the 2015 and 2016 drought in Mozambique, the amount of time women spent fetching water more than doubled.[15] During droughts, girls are often taken out of school, as they are needed to join women in water-collection activities. Unfortunately, climate change is already increasing the frequency and severity of droughts. According to UNICEF, by 2025 half of the world's population could be living in areas facing water scarcity. As a result, UNICEF estimates that by 2030, lack of access to water may result in the displacement of an estimated 700 million people.[16]

Many insects carrying viruses, such as mosquitos, breed in and near water.[17] Women's close association with water sources such as wells, rivers, and ponds makes them more at risk of vector-borne diseases such as malaria, Zika, dengue, and other viruses.[18]

Fuelwood collection is another chore most frequently assigned to women worldwide.[19] Over one-third of the world's population,

2.6 billion people, depend on wood and traditional fuels for cooking.[20] Often this fuelwood is collected from the forest floor. As forests disappear, women are walking farther and spending more of their day collecting fuelwood.[21] In one study in Tanzania, for example, 70 percent of interviewees reported having to walk farther to find fuelwood as a coping strategy for scarcity, and the same is true in parts of Kenya.[22]

Along with being tasked with fetching water and fuelwood, in some countries women are often responsible for foraging for wild fruits, vegetables, leaves, seeds, honey, eggs, and insects. In the last forty years, we have lost 60 percent of global population of mammals, birds, fish, reptiles, and amphibians.[23] When wildlife as a food source becomes scarcer, there is greater reliance on women to provide the family with enough daily calories through these other wild foods.

Preparing food is another responsibility most frequently assigned to women. While cooking over open fires in poorly ventilated spaces, women inhale a lot of smoke, and therefore disproportionately suffer from respiratory illnesses.[24] To date, about 4 million deaths worldwide are a result of household air pollution.[25] A study by the World Health Organization found that women and children account for 60 percent of all premature deaths attributed to household air pollution.[26] In addition, studies have confirmed that long-term exposure to air pollution is linked to an increased risk of dying from COVID-19.[27]

How does this relate to climate change and resource depletion? When fuelwood is scarce, there is a shift to alternative fuels such as burning dung and other products often contaminated with crop residues and pesticides, which release even more toxins than fuelwood when burned. This is increasing the rate of respiratory infections, lung cancer, and other pulmonary diseases in women.[28] Women also suffer from other pollutants. For example, in East Asia, women who engage in near-shore fishing handle plastics that get caught in nets, exposing them to the hazardous chemicals, which affects their reproductive health.[29]

Poverty, working in tandem with women's roles as caregivers, water carriers, fuelwood seekers, and food preparers, increases their vulnerability to the climate and biodiversity crises. Additionally, there are other cultural barriers that confine and limit women living within patriarchal societies and thus amplify the impacts of climate change. For example, in many countries there are differences in what kind of skills women and men are taught. Globally, men are often trained in and practice lifesaving skills such as swimming and tree climbing that are not considered "appropriate" for women. There are also other barriers that prevent women from learning these skills. Faith Milkah Muniale tells me, "Swimming and survival techniques was one of the courses that I had to do for my wildlife degree. Unlike many other communities who live next to rivers and lakes, I didn't swim until I was nineteen. I had never been in water. Sometimes when it was time for swimming class, and I was on my period, I didn't know how to manage the situation. The teacher was male. In the village, you don't talk about these personal things. So, I just missed the class."

According to the 2019 Lloyd's Register Foundation World Risk Poll, based on over 150,000 interviews by Gallup in 142 countries and territories, over half of the world's people cannot swim unassisted. Over two-thirds of women worldwide cannot swim, compared with about 40 percent of men. This gap is wider in the Global South.[30] Within countries, the inability to swim is related to a wider gender gap and lower income. In the lowest income group, 61 percent of men and 85 percent of women say that they can't swim.

In Sri Lanka, tree climbing and swimming were skills that helped men to survive the 2004 tsunami. In an Oxfam report, Shanthi Sivasanan, an Oxfam programme assistant, is quoted saying, "Many men climbed trees to escape the water—it was something they had done many times before to pick fruit and while playing—yet women had never done this before and so didn't do it."[31] In a study conducted by Oxfam in the Aceh Besar district of Indonesia, inability to swim was reported as one of the main reasons that only 189 of 676 survivors of the 2004 tsunami were female.[32]

Women sometimes do not have access to information that can provide an early warning of approaching natural and anthropogenic hazard. Sometimes women cannot read warnings because they are not allowed in public spaces where notices are posted, or, because of illiteracy, they are unable to read them. For example, before the 1991 cyclone in Bangladesh, signs about the approaching cyclone were posted in public places not frequented by women. Globally, women have less access to radios and phones. In 2019, less than half of women worldwide had access to the internet, compared with 58 percent of men.[33] According to the World Economic Forum, the digital gender divide is widest in South Asia.[34]

In many cultures, women are expected to wear certain clothing that can make them more vulnerable to extreme climatic events. Long skirts and wraps of material can restrict movement when women try to run and can weigh them down in water. Behavioral taboos can also prevent women from fleeing. In some countries, there are rules against women leaving their homes without the consent of their husband, brother, or father. During the 2005 earthquake in Pakistan, for example, because of the requirement to cover themselves before leaving the house, women did not leave buildings in time even when the structures were shaking.[35]

Worldwide, every year, weather-related events are disrupting the education of about 37.5 million students.[36] This is partly because schools become damaged or the transportation to schools is shut down. In South Asia, for example, in 2017 after one of the worst floods in the region, eighteen thousand schools were closed. Students' education is also disrupted because parents keep their children out of school to help with the increased work burden arising from a flood or drought. In this way, girls miss out on schooling more than boys since girls are more often kept at home to help with chores. According to a 2008 case study, for example, 70 percent of students taken out of school during droughts in Botswana were girls.[37] Girls suffer from missed schooling also in the long term because even once the immediate emergency is over, they are less likely to return to school than boys.[38]

Clearly, it is not our biology that makes us more vulnerable to severe climatic events or the degradation and pollution of our planet. Women's vulnerability is socially constructed. By taking away rights, forcing women into poverty, piling on chores and responsibilities that take time away from education and employment, restricting attire, movement, and information, women are pushed into situations where they become vulnerable.

In this chapter, I aimed to make the case that women are disproportionately impacted by climate change and nature's destruction, by bringing together research from a variety of sources from across the world, but the truth is that none of this is new. In an article written almost a decade ago entitled "Disasters Are Gendered: What's New?," Joni Seager had already pointed to hundreds of reports providing evidence that disasters were gendered, from immediate impacts to the subsequent social and economic impacts related to post-disaster reconstruction.[39]

Part of the problem is the conflation of the word "gender" and "women." Gender concerns are often seen as a women's issue rather than a crosscutting one. The most common ministerial positions filled by women worldwide, for example, often encompass gender affairs (e.g., Ministry of Gender Equality and Family, Ministry of Gender, Children and Social Protection, or the Ministry of Gender, Family and Children).[40] Another problem is that in nature conservation work, many people do not know where to even begin to address or "mainstream" gender.

An important cause of the lack of attention to gender is that, as Chantal de Jonge Oudraat and Michael E. Brown write, "the male-dominated climate-studies community sees climate change primarily as a scientific, technical, and economic problem."[41] But we need to address our planetary polycrisis from all sides. We need scientific, technical, economic solutions, but they are never going to be sustainable if the gender issues are not front and center as well. Climate change is a social justice and feminist crisis as well as an environmental one.

Fortunately, there is growing awareness of the inequalities in the impacts of climate change. In fact, many people are now objecting to the use of the term "natural disasters" to describe the catastrophic social effects of floods, fires, tsunamis, and droughts. To understand why, I reached out to Kristina Peterson, creator and founding board member of the National Hazards Mitigation Association and the Gender and Disaster Network. She tells me that "literature that embraces gender and justice does not call disasters 'natural.' There are natural hazards that turn into disasters via socially constructed violence and vulnerability," and "the misnomer that disasters can be natural is a false claim that feeds the colonial-economic exploitative narrative." Not everyone agrees with this terminology, however. As climate change wreaks havoc on our weather systems, fires and flooding are not only disastrous to people, but wildlife, forests, and whole ecosystems as well. They are indeed disasters affecting nature, but their root causes are not natural.

Getting this terminology right is important because it shifts the focus back to what humans are doing to create disasters and because it gives us agency to do something about it. While the work of changing patriarchal systems that cause inequalities may seem daunting, knowing that women's vulnerability is a result of these systems and not our biology at least makes it plain that decreasing the impacts of our planetary polycrisis on women is fully within our control.

# 4

# BRAIDED INEQUITIES AND THE ROOTS OF CLIMATE CHANGE

> Patriarchy isn't just entwined with the systems of colonization, white supremacy, and capitalism. Colonization, white supremacy, and capitalism need patriarchy to work.
>
> —*Jihan Gearon*[1]

I AM IN GUINEA AGAIN AND STANDING IN THE SAME PLACE I STOOD two decades ago, looking out across a landscape where chimpanzees were once abundant, their pant-hoots echoing through the forest at dawn, and their nests folded into the forks of trees at dusk. In the heat of the day, young chimpanzees played in the shade of towering trees, while their parents groomed each other. But today, I look out across a bleeding red earth. Nothing is left except a vast plane of exposed dirt with scars, scrapes, and roads gouged into its surface from machines carving out bauxite and hauling it away to other countries in unfathomable quantities to make aluminum for high-end cars, planes, rocket ships, and golf clubs. A plethora of companies from all over the world—Russia, Australia, the United States, United Arab Emirates, and China—are here dynamiting and digging Guinea's earth and transporting it across the Atlantic Ocean to foreign nations, leaving a moonscape where there was once an abundance of life.

I am with my friend and filmmaker Kalyanee Mam along with Mamadou Diawara, director of the national NGO Guinée Écologie, to learn more about the impacts of the mining on wildlife and people. We have just visited a bauxite mining concession. The area is a sea of treeless red earth. There is very little activity today, however. We learn that the workers are on strike to protest their cruel and abhorrent working conditions in the mine. We hear that workers are beaten, and that when a mining truck allegedly hit and killed a person walking along the road, the miners stopped working and took to the streets in protest. This is not new. For years now the people of Boké have marched against the bauxite dust that choked the air and polluted the waterways. They have protested low wages. They threw rocks at the government buildings, because despite all the money Guinea was earning from selling their land, there was still no electricity or running water in most houses. Several months before our arrival, there were similar demonstrations and several protesters were shot and killed.

We leave the site and take a small dirt road toward a village next to the concession, where we meet with several elders of the village and ask questions about how the adjacent mining concession has affected them. One of the men, Amirou Diallo, takes us to see the river where the people of Hafia village used to get drinking water. What now remains of the river is a small depression in the earth filled with water. Livestock come and go to drink from it. A mother soaps up her toddler next to the pool, and then pours water over her head to rinse her.

Amirou says that people in the village have been getting sick from the water since the mining began nearby. He complained to the mining company, and he says that it built a well for the village. He shows us the well, and pumps some water, which has a rainbow sheen like that of gasoline on a wet surface. He says when they boil it, the liquid turns black. "We used to be in the gallery forest of the large rivers with the chimpanzees," Amirou says. "Everything has disappeared, particularly the chimpanzees. And the fish we used to catch, there are none left. The watercourses we used before all dried up. The rocks that retained the water before were dynamited. The large trees were uprooted, and the mining debris flows towards the watercourses. Everything the mine discharges goes down to accumulate in the water that then dries up. We are 350 people. The suffering has become unbearable."

I struggle with the word "development." Industrial mining is not "development"—not for the village of Hafia in Guinea. So often the rampant exploitation of nature does not benefit those most affected. This is true for the climate crisis as well. Those most affected have benefited the least from the burning of fossil fuels.

The level of greenhouse gases first started to rise during the Industrial Revolution, which began in the late 1700s in England and spread to Belgium, Germany, France, Japan, and the United States by the 1800s. To become an industrialized nation, countries needed capital to invest up front in building machines, from which they would later reap the benefits. Not all countries had such capital, but those that did became more industrialized and, consequentially, increasingly wealthy. This permitted them to expand their military power and colonize more countries, which increased their wealth even further.

As their economies grew, industrialized countries needed more raw materials such as rubber, cotton, metals, and sugar, so they began sourcing these from poorer countries, where they dug mines, razed forests, created immense plantations, and imported people as slaves. A tremendous shift in power and wealth took place across the world as a result. Many countries supplying these raw materials, instead of growing diverse crops to feed people, began to grow cash monocultures that fed the demand from foreign countries. As a result, the economies of source countries became less diverse and more dependent on export to richer countries.

Vast areas of countries, whether set aside for plantations or mining concessions, were owned, managed, and controlled by foreign nations. Profits and benefits seldom made it back to those people in countries displaced by the industrial exploitation of their land and resources. In many places like Guinea, this is still the situation today.

Since the beginning of the industrial era, humans have caused, and continue to cause, so much destruction to the planet that scientists have created a name for this period: the Anthropocene. The Greek word *anthropo* means "human," and *cene* means "new." First used in the year 2000 by Eugene Stoermer and Paul Crutzen, the term has now appeared in over 1,300 scientific publications worldwide, yet it is still considered unofficial. The organization that names and defines epochs, the International Union of Geological Sciences, has not formally adopted the term since we have not yet answered the chilling question of whether humans have changed the Earth to such extremes that we can see it in rock strata.[2]

But is it fair to say that such destruction is created by "humans," as if we are all equally responsible? Indigenous, Black, and feminist scholars have pushed back against the accuracy of the term "Anthropocene" because it implies that the climate and biodiversity crises are a result of the actions of all humans, rather than "a minority of colonialists, capitalists, and patriarchs."[3]

When we disaggregate data by gender, geography, race, and other factors, it of course becomes clear that not all countries or people have contributed or continue to contribute to the same degree. For example, Asia contributes disproportionately to ocean plastic,[4] while Australia and the United States are the top single use plastic producers per capita.[5] Deforestation rates also vary widely between countries and over time. For example, from 2002 to 2022, the decrease in total area of humid primary forest in Brazil was 8.6 percent, while in Gabon, it only 1.2 percent.[6]

With regards to climate change specifically, not only are women in the Global South disproportionately impacted by climate change, but they are also the least responsible for creating it.[7] As Anne Karpf, author of *How Women Can Save the Planet*, warns: "Beware the climate 'we.'"[8]

Historically, the twenty-three richest countries on the planet are responsible for 50 percent of all the greenhouse gases in the last 170 years. These same countries make up only 12 percent of the world's population. The top five emitters have historically been the United States, Germany, France, the United Kingdom, and Japan. The 150-plus poorest countries in the world who make up the remaining 88 percent of the world's population have historically emitted the other half of the greenhouse gases that are in the atmosphere today.[9] More recently, however, several other countries are catching up in terms of emissions. For example, China has historically been responsible for only 19 percent of emissions, but today is at the top of the list of the worst emitters. In fact, China is now responsible for almost one-third of all annual emissions. The top ten emitters in 2022 were China (30 percent), the U.S. (13 percent), India (7 percent), Russia (5 percent), Japan (3 percent), Germany (2 percent), Iran (2 percent), South Korea (2 percent), Indonesia (2 percent) and Saudi Arabia (2 percent).[10]

Not only are a handful of countries responsible for most greenhouse gas emissions, but when we look at which *companies* are responsible for the emissions, we can narrow things down further. The Carbon Majors Database revealed that in 2015, fossil fuel industry companies were responsible for 91 percent of all global industrial greenhouse gases that year, and about 70 percent of all anthropogenic greenhouse gas emissions.[11]

Human-induced climate change was first officially recognized in 1988, at a meeting between the United Nations Environment

Programme and the World Meteorological Organization. At this same meeting, the Intergovernmental Panel on Climate Change (IPCC) was created. Since that moment when climate change was officially recognized, the fossil fuel industry has *doubled* its contribution to global warming.[12]

We can break this down even further. A total of twenty-five companies account for 51 percent of global industrial greenhouse gas emissions, and 100 producers account for 71 percent of global industrial greenhouse gas emissions.[13] Katharine Hayhoe, chief scientist at The Nature Conservancy (TNC), has shared the helpful visualization that you could fit all the CEOs of these companies in three buses.[14] Notably, only 7 percent of the CEOs on those buses would be women.[15] Women have not been equally responsible for the leadership that has led us into this planetary polycrisis.

Even though those companies are producers of greenhouse gas emissions, we are all responsible when we use fuel, drive cars, heat our homes, or ride in airplanes. But again, "we" is not a uniform group. If we look at greenhouse gas emissions at the individual level, those whose income is in the top 10 percent of global income are responsible for 40 to 60 percent of total greenhouse gas emissions.[16] An analysis in 2015 found that the share of total carbon emissions for the people in the bottom 50 percent of income globally was half that of the people in the top 1 percent. In addition, the authors of the study note that "since the bottom 50 percent has 50 times more people in it, the average per capita consumption emissions linked to the top 1 percent in 2015 were over 100 times greater than the average per capita consumption emissions of the poorest half of the world's population."[17] Of those top earners, only 4.5 percent are women.[18] From this perspective, it appears that women have not been equally responsible for greenhouse gas emissions either. (Although of course, if women are in families of the top 1 percent income, they may have just as high per capita consumption emissions).

In addition, some have been quick to point out that, in the United States, women are on average making more purchases than men, which could mean that women generally have greater per capita consumption emissions.[19] However, purchases women make are often not only for personal needs but for household ones, such as groceries. In an article in *Forbes*, Bridget Brennan writes, "In virtually every society in the world, women have primary care-giving responsibilities for both children and the elderly (and often, just about everybody else in-between). In this primary caregiving role, women find themselves buying on behalf of everyone else in their lives."[20]

Curious to hear how women on the front lines of climate change feel about the unequal responsibility of climate change between nations, I ask Madina Hussein, founder of Global Nature Conservation, and an Indigenous conservationist in Kenya. She says, "We realize that people at the grassroot, especially women, vulnerable groups, and young people, are greatly disproportionately impacted by climate change, yet they contribute the least for climate change to happen. It's very unfair. It doesn't make sense."

This was the same sentiment many countries from the Global South voiced at the Climate Conference of the Parties in Egypt 2022, which Madina attended. Countries have been pushing back against the global "we." During climate change negotiations, many of the 150-plus poorest countries think that the 23 wealthiest countries should pay a little more in building climate change resiliency and mitigating and repairing damages. At COP27, the parties made a historic decision to establish a "loss and damage" fund, which would especially support those nations most vulnerable to the climate crisis.[21] Hopefully, these funds will be distributed to countries like Pakistan, which has already had to spend $30 billion because of catastrophic flooding, even though it emits less than 1 percent of all emissions annually.[22] Details still need to be resolved, including who will pay into the fund, which countries will benefit, where the money will come from, and who will be in a decision-making position to determine what it is spent on. It will be critical to ensure that funding within recipient countries flows toward the most impacted and marginalized groups, including women. People like Madina are working to make sure this happens.

Along with a lack of sufficient attention to the unequal impacts of climate change on women, there has been little focus on identifying who contributed to climate change in the first place, who is benefiting from fossil fuel consumption, and who is being compensated for damages or supported for mitigation. One of the reasons we are not hearing the voices of women is that they continue to be excluded from national leadership as well as decision-making forums about climate change and the future of our planet. According to the Pew Research Center, as of March 2023, women are the heads of government in only 13 out of the 193 member states of the United Nations (7 percent). Less than one-third of UN countries have ever had a woman leader.[23] Worldwide, women make up only 21 percent of government ministers. In some countries such as Armenia, Azerbaijan, Brunei Darussalam, Papua New Guinea, Saudi Arabia, Thailand, and Yemen, there are no female ministers. In 2020, globally, only 15 percent of environmental ministers were women.[24]

This inequality also exists within international conservation NGOs working to protect nature. White men make up 75 percent of CEO positions at these organizations. Studies have found that in general, in the conservation sector, women hold more positions that require "soft skills and administrative roles," whereas men fill positions that are perceived as more "risk taking," especially fieldwork.[25]

Another way in which women's voices in natural resource management are being limited is their lack of representation in scientific journal publishing. To look at this in more detail, Robyn James of TNC searched the "Web of Science" database for all papers between 1968 and 2019 and recorded those that had at least one author from TNC. The resulting paper, on which Robyn was lead author, found that women made up just over one-third of these authors and less than one-third of lead authors.[26] Women from the Global South made up less than 3 percent of all TNC authorships.

Robyn's paper also found that in a 2018 article in *Nature Ecology & Evolution*, only three out of "100 articles every ecologist should read" were authored by women. In a 2021 survey of the "top-publishing authors" between 1945 and 2019 in thirteen leading journals in ecology, evolution, and conservation, only 11 percent were women (and more than 75 percent of the top-publishing authors were from the United States, the United Kingdom, Australia, Germany, and Canada).[27] In 2021, Reuters published a "hot list" that ranked 1,000 of the most influential climate scientists, based on a combination of three rankings. Out of the 1,000, only 122 were women.[28] Out of the top 100, only 8 were women.[29] Only 111 of the climate scientists were from the Global South.

When women do publish about climate change, they receive more online harassment and threats than men. A shocking report by Global Witness reveals that overall, 39 percent of the scientists the organization surveyed had received online hate as a result of their work on climate change. There were gendered differences too. Global Witness reports, "Their sex or gender was targeted a great deal or a fair amount for 34 percent of affected women and only 3 percent of affected men. Similarly, women were three times as likely as men to receive a great deal or a fair amount of harassment on the basis of their age (17 to 5 percent) and received more threats of sexual violence (13 percent of affected women) and physical attacks than men. Almost a fifth (19 percent) of women and

16 percent of men who had faced harassment had received threats of physical violence."[30]

One of the most critical ways that women can have their voices heard about climate and biodiversity issues is at the Conference of the Parties. The COP for the United Nations Framework Convention on Climate Change meets every year to assess progress in dealing with climate change. A separate COP, held for the Convention on Biological Diversity, meets every two years. In the 2022 Climate Change COP in Egypt (called COP27 because it is the twenty-seventh meeting), the proportion of women party delegates was just over one-third.[31] It is not only the number of women, however, that is important, but also the proportion of speaking time. At COP26, women made up just over one-quarter of the total speaking time.[32] In other words, we were listening to men for almost three-quarters of the time.

Instead of saying that women were underrepresented, some prefer to say that this is an "overrepresentation of men."[33] This overrepresentation exists not only at the head of state level but throughout different levels of climate change negotiations as well. The first assessment report that the IPCC published in 1990 had about a hundred authors, of which 90 percent were men and more than 80 percent were from institutions in the Global North. Almost three decades later things have improved, although not enough. In the IPCC report published in 2023 with more than seven hundred authors, around 70 percent were men and 60 percent from the Global North.[34]

A 2022 paper in *Nature* highlights the challenges for women within the IPCC. Data from surveys reveal that while working with the IPCC, more than one-third of women had observed someone else taking credit for a woman's idea (38 percent of female respondents), or had seen a woman ignored (52 percent) or patronized (41 percent). Also, a notable number of women had experienced or observed sexual harassment (8 percent and 11.5 percent, respectively).[35] When asked about hurdles to their participation, almost twice as many women (44 percent) as men (24 percent) reported childcare responsibilities as a barrier.[36]

In a 2023 article, Robyn James and colleagues looked at how gender biases are holding women back in their conservation careers, specifically at TNC.[37] They found that under half of the women working at the organization felt that they had influence in affecting research and conservation priorities within the organization, compared with almost two-thirds of men. Before joining TNC, 29 percent of women felt that their gender had been a factor causing them to miss out on a raise or promotion, compared with 4 percent of men. Fifteen percent of women reported to have experienced sexual harassment at conservation-related conferences or important meetings, compared with 2 percent of men.

Both the IPCC and TNC analyses reveal that men and women perceive gender biases differently. Robyn James writes that in conservation organizations, "there is often a reluctance by men to acknowledge or reflect on gender bias in their workplaces."[38] Similarly, the 2022 *Nature* study about the IPCC found that, compared with women, men consistently underreported observations of women being the targets of sexist behavior. Without men noticing and recognizing its existence, gender bias is more challenging to address. In an article for the Wilson Center, Chantal de Jonge Oudraat and Michael E. Brown go even further to suggest that gender issues are intentionally disregarded since "ignoring these core problems allows men to circumvent their own responsibilities."[39]

Another huge barrier for anyone's participation at any level of work to address biodiversity and climate crises is funding. Funding is key to having a voice. Funding is needed for research. To share ideas and knowledge, funding is also needed for flights to meetings, hotel lodging for conferences, and to publish in journals, launch websites, access the internet, and cover childcare costs.

*Our Voices, Our Environment: The State of Funding for Women's*

*Environmental Action*, an online report published by the Global Greengrants Fund, Prospera International Network of Women's Funds, and Global Alliance for Green and Gender Action, was, in its authors' words, "the first-ever benchmarking of philanthropic funding in support of women and the environment."[40] Shockingly, the report found that funding in 2014 for women and the environment (funding for the environment that specifies support for women, girls, or gender equality as well) represented less than 0.1 percent of the number of foundation grants and only 0.2 percent of the overall amount of foundation grant dollars. The same report found that less than one-third of donors gave more than one consecutive grant, and only eight donors gave more than $1 million to supporting projects with an explicit focus on women and the environment. Much of this funding was for "agricultural development and food security research to benefit smallholder farmers." The report also pointed out that even though the grants were going to the Global South, they generally went to research institutions and universities and not directly to local women's organizations. The conservation news platform Mongabay also reports that "of all the philanthropic funding to tackle climate change, 90 percent goes to organizations led by white people, and 80 percent to organizations led by men."[41]

Underrepresentation in government leadership, conservation organizational leadership, scientific publications, climate meetings—specifically the Climate COPs with respect to speaking time—and the paucity of funding bind together to form one huge barrier to the equal inclusion of women's voices, especially voices of women from the Global South and women of color. Excluding women misses a huge opportunity to truly understand the complexity and range of climate change and biodiversity loss impacts on nature and on ourselves, and it limits us in finding effective ways to address our planetary polycrisis.

There is a logical disconnect here. Women are disproportionately affected by climate change. Women are relegated to prescribed roles and forced into cycles of poverty that leave them most dependent on nature. Women, especially women in the Global South, have contributed the least to greenhouse gases and benefited the least from the burning of fossil fuels. Yet women are also underrepresented in positions that can influence decisions about nature. As we will see in the next chapter, all evidence to date shows that when women are empowered, when they do have equal participation and lead in decision-making about nature and climate, both people and the environment thrive.

# PART II

# TRUNK AND BRANCHES

Trees do not sense which way is up; rather, their growth follows the direction from which light comes. This phenomenon is called phototropism, which means bending toward the light.

—*C. Claiborne Ray*[1]

# 5

# MOTHER NATURE, MOTHER EARTH

> The war against the Earth began with this idea of separateness. Its contemporary seeds were sown when the living Earth was transformed into dead matter to facilitate the industrial revolution. Monocultures replaced diversity. "Raw materials" and "dead matter" replaced a vibrant Earth. Terra Nullius (the empty land, ready for occupation regardless of the presence of Indigenous peoples) replaced Terra Madre (Mother Earth).
>
> —*Vandana Shiva*[1]

EARTH, IN ALL ITS DIVINE, STARTLING BEAUTY, WITH RISING MOUNTAINS, deep canyons, flowing rivers, expanses of humid tropical rainforests, fertile soils, and an abundance of colorful wildlife, has been personified as a woman throughout time and across cultures. The word "mother" has been used to represent weather, oceans, land, and the cycles of seasons. She can create life and take it away as well. She is caring, ruthless, powerful, and gentle. She is all of this. At least, she used to be. In the same way that patriarchy, capitalism, and colonization joined forces to oppress, exploit, and devalue women, these same forces used the earth for the gains of a few and the suffering of many. Just as the history of the treatment of women and the earth have been intertwined, so are the fight for the rights of both.

In Inca mythology in the Colombian Andes, Pachamama, "Earth Mother," was the source of life, weather, water, food, and animals, and was responsible for fertility, planting, and harvests. All forms of life on earth are her children.[2] In many Indigenous cultures from the Pacific Islands, Papahānaumoku (often called "Papa" for short) is the earth goddess. According to goddess scholar Patricia Monaghan, Papa's lover was the sky god, Rangi, and they were "in perpetual intercourse." Some of the gods created by the union between Papa and Rangi had to live in the darkness between them. One day, they decided to separate their parents. Once apart, they missed each other, and today Rangi's tears fall down to Papa and her warm mist rises toward him.[3]

For the Bribri people in Costa Rica, the cacao tree is believed to be a woman who was turned into a tree by God. Only women are allowed to collect the cacao branches and use them as firewood, and only women are allowed to prepare and serve the sacred drink made from cacao.[4]

In ancient Egypt, a goddess named Isis was the giver of life. In pre-Christian Celtic mythology, Nantosuelta was the goddess of nature. Jörð was a pre-Christian, female personification of the earth. Máttárahkká is the "earth mother" and goddess of the Sámi, Indigenous people of Europe. According to Astri Dankertsen, associate professor in sociology at Nord University in Norway, Máttárahkká is the "mother of the tribe, goddess of women and children, and it is she who gives humans their bodies. She is also, together with Sáráhkká, one of her three daughters, the goddess of fertility, menstruation, love, human sexuality, pregnancy, and childbirth."[5]

Called by many different names such as "heroines" and "spirits," goddesses were unlike the familiar female figures worshipped by many religions today, who are pious and dutiful, with eyes downcast in obedience and reverence. Many ancient goddesses did not embody virtuous behaviors on which humans were expected to model themselves. They were often promiscuous, and became "impregnated by wind or ocean waves, by snakes or fiery flames, or simply by their own desire."[6] They were not all mothers either. Some had hundreds of children, while others had none. They were young and old. They were warriors. They were huntresses. These goddesses demonstrated the infinite possibilities of being female, and refused to accept any barriers to or beliefs about what a woman could do, could not do, or should not do.

Archaeologist and prehistorian Nicholas Conard holds a life-size replica of the Hohle Fels figurine close to the camera on Zoom so that I can see her more clearly. While digging near the mouth of a cave in the German Alps in 2008, under about three feet of earth, he found a tiny female figurine carved out of a mammoth tusk. She is about six centimeters long and weighs less than one ounce. The carving features distinct fingers of each hand—four on one hand, five on the other. She has "upright, oversized breasts and massive shoulders relative to the flat stomach."[7] Her legs are short and pointed at the ends. Her buttocks and genitals are carved in more detail than other parts of the body. The Hohle Fels is one of the oldest pieces of art representing the human form, having been carved by humans between 35,000 and 40,000 years ago.[8]

"Nothing is there by chance," Nicholas tells me, because ivory is hard to work. Whatever features this figurine possesses must have been carved intentionally. Because the Hohle Fels has no feet, she probably does not represent movement. She has no head, so it is more likely that she represents a concept or is a symbol of something rather than a depiction of a specific person. She has a ring where her head should be, so it is possible that she was worn as a pendant. One curious thing about her is the distinct lines carved across her stomach, as if to indicate loose rolls of skin. Many say that these lines are to indicate she is not youthful. It is likely that she represents an older woman.

Although the Hohle Fels figurine is the oldest, many other ancient female figurines have been found around the world, and of particular interest are those discovered in a more recent archaeological site in central Turkey, called Çatalhöyük, where humans lived between about 9,400 and 5,900 years ago. What is interesting here is that although many female figurines were uncovered, no male figurines have been found that are as "elaborate or highly crafted."[9] One female figurine from Çatalhöyük, known as the "Seated Woman," even appears to be sitting on a throne. About eight thousand years old and made of baked clay, she is nude, with large, sagging breasts, and has creases carved into her knees. She is seated between what archaeologists hypothesize are armrests, with two large felines on either side of her. Her appearance is authoritative and confident.

Many people, especially adherents of the Goddess movement, have taken these female figurines to be evidence that humans once lived in matriarchies. While it is unlikely that humans evolved from a universal matriarchy, new techniques to analyze the DNA of skeletons have upended many of our early assumptions about gender roles. Many skeletons once believed to be male turned out to be female. For example, in 2020 a discovery of ancient bones next to hunting tools made headlines across the world. High in the Andes mountains, Randy Haas of the University of California-Davis uncovered a nine-thousand-year-old skeleton in a burial pit with "a tool kit of 20 stone projectile points and blades stacked neatly by the person's side."[10] At first, he assumed that the skeleton was that of an important man—a chief or hunter—yet his analysis revealed the skeleton to be a woman. Randy then became curious to see if perhaps archaeologists had been making incorrect assumptions about the gender of hunters all along and so conducted a meta-analysis of other burial sites throughout the Americas, finding that many of the hunters assumed to be male were in fact female.[11]

Vivek Venkataraman, assistant professor of anthropology and archaeology at the University of Calgary, points out that the myth of "Man the Hunter" as a driving force for the evolution of humans was already debunked half a century ago. In an article in *The Conversation* titled "Ancient Men Were Hunters and Women Were Gatherers. Right? Wrong," Vivek explains out that back in 1966, at a meeting in Chicago that brought together seventy-five anthropologists (93 percent of whom were men) working at sites across the globe, the myth of "Man the Hunter" was already rejected based on their cumulative evidence.[12] Yet it is a myth that popular culture still finds it hard to let go of.

A July 2023 CNN article reported on the discovery that the "Ivory Man"—a five-thousand-year-old skeleton discovered in 2008 in Spain, buried with an ivory tusk, crystal dagger, and other objects of wealth—was in fact "Ivory Woman."[13] She was believed to be a "revered" leader and to have come to that position by merit, not inheritance. Similar studies have shattered the myth of "Man the Warrior,"[14] "Man the Cave Painter,"[15] and "Man the Toolmaker."[16] So what do these findings together reveal? They demonstrate that our assumptions about gender roles in our ancestors have turned out to be incorrect and that there was much more diversity in gender roles than we ever expected.

If that is true, why, and how, did patriarchy come to be the most common social system for humans across the planet? To answer this question, I reach out to Margaret W. Conkey, professor emerita of anthropology at the University of California, Berkeley. Meg was one of the first anthropologists to apply a feminist lens to archeology. Her work has revealed that the superimposition of modern cultural values on the past can influence interpretation of evidence, often incorrectly. As a result, she has challenged many assumptions about gendered roles from Paleolithic times.

When I write to Meg, I learn that we live only a few blocks apart, so we arrange to meet for coffee. In preparation for our meeting, she warns me that I may be disappointed by her answer, but adds mysteriously that she considers the truth about our prehistoric roots to be one that is still very hopeful. What I am looking for in the past is certainty, but Meg explains that ambiguity characterizes what archaeology is all about. "We don't nor will we ever really 'know for sure.' How could we? People are complex," she writes to me before our meeting.

I meet with Meg in between rain showers in what has seemed like a relentless series of storms battering the California coast during the spring of 2023. Over coffee, Meg explains the reality is that the way our earliest human ancestors lived and the structure of the relationships between them was varied. We survived across differing landscapes with different threats and pressures, and our human ancestors experimented and adapted. As a result, we had many different types of social systems.

For those systems that were originally more egalitarian in terms of gender, some scholars have historically pointed to the invention of agriculture as one of the underlying drivers of patriarchy. Farming, it has been argued, required staying put, and this new sedentary lifestyle could have been at the root of other changes that followed. Before agriculture, we were mobile. We lived in family units within larger groups. We hunted, fished, gathered, and shared food cooperatively with unrelated individuals. These multifamily groups are believed to have had fluid social networks.[17] Sometimes families lived with the mother's family, sometimes with the father's family. According to Sarah Blaffer Hrdy, anthropologist, primatologist, and evolutionary theorist, boundaries between groups were "porous."[18] In her book *Mothers and Others: The Evolutionary Origins of Mutual Understanding*, Sarah presents abundant and convincing evidence that to provide enough calories for human babies who were slow to mature and establish their independence, mothers in the Pleistocene needed help. As a result, they relied on others—grandmothers, sisters, aunts, older children, uncles, grandfathers, and other men—to care for their infants, something called "alloparenting." This kind of support not only increased the survival chances of the infant but allowed the mother to have more children.

One theory regarding what drove patriarchy's rise is that once humans started cultivating their own food, stopped their nomadic lifestyle, and began to own things—especially land—it became important for sons, brothers, and fathers to stick together so they could defend property more effectively. Women were then more frequently obliged to live in the house of their husband, and the boundaries between families became less fluid.[19] Critically, to ensure men were passing property and possessions on to their biological children and not someone else's, men needed to find a way of controlling women's reproduction. The problem for women was that living in the husband's home isolated them from the support they had received from their mothers, grandmothers, aunts, and sisters—a support system not only for childrearing but for everything, including standing up to the power of men. Could it be that this isolation of women and the breaking of their support networks is what permitted patriarchy to take hold, and what made it very difficult for women to regain power?

Earth and nature were once seen as a living, breathing woman, but around 2,500 years ago, with the spread of monotheism and patriarchy, many goddesses were erased and worship of them forbidden. Where goddesses did persist, they were often disguised, absorbed into colonizers' own religious stories, or worshipped privately. For example, in the Amazon, when Spanish colonizers forced Roman Catholicism upon Indigenous cultures, the goddess Pachamama endured, but her character was often combined with the Virgin Mary's.

For Midori Nicolson, Musgamagw Dzawada̱'enux̱w from

Gwayi, Kingcome Inlet, it was only in her adulthood that she learned of how the origin story of her people had been influenced by colonialism. For thousands of years, the Kwakwaka'wakw First Nations have lived in this inlet on the mainland of British Columbia opposite Vancouver Island. Today, Midori lives alongside the glacier-fed Kingcome River, surrounded by high mountains. She works within her community as a biologist and fisheries manager, with a focus on ensuring the inclusion of women's voices in the stewardship of nature and the land.

When we speak, Midori describes her discovery of how a key female figure in her people's origin story, named Hayatliligas, had been somewhat erased:

> The story I had been told my whole life of Kawadilikala and Kwalili, two wolf brothers who come down the watershed from the glacier to found two major villages in our territory, had only a slight mention of the sister, Hayatliligas, who they leave behind up in the watershed to "watch over the land." When Kawadilikala tested his supernatural power against Hayatliligas by throwing a quartz crystal back and forth, he could not defeat her and ended the contest by throwing the quartz at a mountain. She then built her house on the upper part of the river Gwa'yi. Hayatliligas's job to watch over the land was actually so very important, yet she had been downplayed or almost erased from the story as a minor role, as the story threads follow the brothers downriver. I wondered what became of her and why didn't we share more of her story anymore? I think that much of our stewardship relationship with our world was a given in those days, and perhaps didn't need defining in the way that it does now, but I think about her all the time and wonder.

The breaking apart of support networks, and the isolation and alienation that follow it, are not just a result of patriarchy, but also of colonization. For many Indigenous groups in North America, matriarchies were only replaced with patriarchal systems within the last two hundred years, and it was because of colonization. During my interview with Midori, she talks about how colonization, genocide, and attempts at assimilation changed the position of women within her culture. She explains to me that women once had a central role, but 150 years of contact changed things:

> As a matriarchal society, women had tremendous decision-making power. When we look at our old stories we can see the roles of these women. We came from a system of many governance seats. Different families had a say in how our culture was structured, and therefore had voice and representation. Then the colonialists boiled it down to one man. There was the fantasy of "here's the chief, his princess daughter, etcetera." All of that was a projection onto our culture, but because our people went to residential school and suffered through colonial genocide, we, in many ways, absorbed that way of thinking. Our own people now believe that power is solely held by a man.
>
> We're struggling because our culture wasn't built this way. Women had tremendous decision-making power culturally, but now, because we're in a recovery of our culture and trying to rebuild it, it's challenging for us, because everywhere we look, in all societies, men rule the world. We've struggled under genocide to reimagine what leadership is, because these guys think it's a crown, or it's a superhero cape that you put on. And it's not.

The removal of children from Indigenous communities across Canada for placement into residential schools had the long-term consequences of altering the very fabric of society and shifting gender roles and relationships. Colonialism worked to erase Indigenous people through genocide, and through the separation of individuals from their ancestral lands, traditions, culture, language, and even from the bones of their ancestors.

Corrina Gould, Ohlone and tribal spokesperson for the Confederated Villages of Lisjan Nation, works in the San Francisco Bay Area, where I live, to return Indigenous land to Indigenous people. When I speak with Corrina on Zoom, she says, "Our territory of Huichin is a place we have never been disconnected from. My mom was born here in Oakland and our family was born in Pleasanton and all around the Bay Area. We have an unbroken tie to the land since the beginning of time."

There is a shopping mall on the edge of Berkeley in Emeryville. I learn from Corrina that this was built on a sacred burial site—a shellmound where Ohlone people buried their ancestors. I also learn from Corrina that the location of the shellmound was no surprise to the developers. In 1909, a man called Nels Nelson had documented and mapped the shellmounds of the Bay Area for the University of California, Berkeley.[20] Corrina says that "when the Emeryville Shellmound was being destroyed, we got phone calls from Native people that were working on the site, saying that there were bodies being pulled out of there. This was our first time dealing with something like this." At a meeting with City of Emeryville leaders where she spoke about the importance of the history of her people, Corrina was asked, "What is it that you want? Do you want medical insurance for your families? Do you want scholarships?" She says, "We didn't know the process of fighting against these things. All we knew was that we wanted to stop it. But we came into the process late. It was already a done deal. There were no laws that really helped us to navigate it, and we didn't have any legal representation. It was just us and a bunch of community members and other tribal people saying that this shouldn't happen. And we lost."

After Corrina lost her fight to protect the Emeryville Shellmound, she started getting more phone calls. "They were digging up bodies in San Jose as they were putting up a condominium. They were digging up bodies in Alameda when they were fixing a pipe leaking under housing the city owned. All these bodies started to come up in a very short amount of time. It was like the ancestors were calling us to do something. So, we answered."

Corrina and her friend Johnella LaRose launched an Indigenous women–led land trust, based in the Bay Area, called the Sogorea Te' Land Trust. What Corrina wants is not just to return Indigenous land to Indigenous people, but to *rematriate* the land. I ask her what "rematriation" means. "It means that it allows Indigenous people to work within the land, and to bring back our culture and our songs without outside interference," she says. "It's a sacred duty for us, but it's also Indigenous women's work. Rematriation is about women's way of doing things. When we look at the colonization of our lands, and things that happened when other people got here, they took away the sacred, they diminished women's roles and took away their power. They began to look at land and waters as commodities, something that were resources, something that they can take and control the same way that they can control women's bodies. Men being charged with the land correlated exactly with what has happened to women and continues to happen to women today."

Patriarchy and colonization work to isolate people from their land, ancestors, culture, and each other. Climate change also works to isolate. Extreme heat and extreme cold keep people inside. The pandemic necessitated social isolation. As climate anxiety rises, so too do levels of loneliness and social isolation.[21] Our overexploitation of nature works to isolate and disconnect as well: deforestation results in isolated forest patches and increases the vulnerability of all life within them.

As patriarchy, colonization, and capitalism spread across the globe, women and the earth followed a similar fate in terms of loss of status and value. The earth in Western symbolism metamorphosed from being a nurturing mother to a sexualized female, vulnerable to exploitation. This evolved further into a view that humans needed to be separated from, and to dominate, nature to thrive. This idea became cemented during the Industrial Revolution (circa 1760–1840), widening the nature-culture binary. Nature was thought of as something not only apart from humans—but as something that stood *in the way of human progress*.

Carolyn Merchant, distinguished professor emerita at the University of California, Berkeley, writes in *The Death of Nature: Women, Ecology, and the Scientific Revolution* that this death in the belief of the earth as a living being was a necessary precursor to the Industrial Revolution, because "as long as the earth was considered to be alive and sensitive, it could be considered a breach of ethical human behavior to carry out destructive acts against it."[22] The earth in the symbolism was purposefully transformed again, this time into dead matter, thus removing any remaining ethical barriers for exploitation.

The environmental and feminist movements have grown like stems and branches of a twisting vine or tree. Sometimes merging, sometimes growing apart. At times they have strengthened each other, yet at other times they have grown distant. Ultimately, they both address similar forces of oppression and exploitation. They share a common goal of dismantling the "status quo." Their shared